Imaginative Ecologies

AF692990

Nature, Culture and Literature

General Editors

Hubert van den Berg (*Palackj University in Olomouc*)
Axel Goodbody (*University of Bath*)
Marcel Wissenburg (*Radboud University Nijmegen*)

Advisory Board

Anabela Carvalho (*University of Minho*)
Heinrich Detering (*University of Gottingen*)
Robert Emmett (*Virginia Tech*)
Adrian Ivakhiv (*University of Vermont*)
Richard Kerridge (*Bath Spa University*)
Michiel Korthals (*Wageningen University*)
John Parham (*University of Worcester*)
Tarla Rai Peterson (*Texas A&M University*)
Ulrike Plath (*Tallinn University*)
Luis Pradanos (*Miami University*)
Kate Rigby (*Bath Spa University*)
Piers Stephens (*University of Georgia*)
Arran Stibbe (*University of Gloucestershire*)
Bronislaw Szerszynski (*University of Lancaster*)
Nina Witoszek (*University of Oslo*)

VOLUME 17

The titles published in this series are listed at *brill.com/ncl*

Imaginative Ecologies

Inspiring Change through the Humanities

Edited by

Diana Villanueva-Romero, Lorraine Kerslake, and Carmen Flys-Junquera

BRILL

LEIDEN | BOSTON

GENERALITAT VALENCIANA

The open access version of the Introduction and Chapter 4 (Author: Lorraine Kerslake) was funded by the Generalitat Valenciana. Conselleria d'Innovació, Universitats, Ciència i Societat Digital. Project number: GV/2020/029 / Angels of the Ecosystem? / ¿Ángeles del ecosistema?

Cover illustration: Vora Mar, by Fernando Franco Sánchez, September 2020.

Library of Congress Cataloging-in-Publication Data

Names: Villanueva-Romero, Diana, editor. | Kerslake, Lorraine, editor. | Flys-Junquera, Carmen, editor.
Title: Imaginative ecologies : inspiring change through the humanities / edited by Diana Villanueva-Romero, Lorraine Kerslake and Carmen Flys-Junquera.
Description: Leiden ; Boston : Brill, 2022. | Series: Nature, culture and literature, 1572-4344 ; vol.17 | Includes index.
Identifiers: LCCN 2021046491 (print) | LCCN 2021046492 (ebook) | ISBN 9789004501263 (hardback) | ISBN 9789004501270 (ebook)
Subjects: LCSH: Human ecology and the humanities. | Literature, Modern–History and criticism. | Arts–History. | Environmental justice.
Classification: LCC GF22 .I43 2022 (print) | LCC GF22 (ebook) | DDC 304.2–dc23/eng/20211029
LC record available at https://lccn.loc.gov/2021046491
LC ebook record available at https://lccn.loc.gov/2021046492

Typeface for the Latin, Greek, and Cyrillic scripts: "Brill". See and download: brill.com/brill-typeface.

ISSN 1572-4344
ISBN 978-90-04-50126-3 (hardback)
ISBN 978-90-04-50127-0 (e-book)

Copyright 2022 by Koninklijke Brill NV, Leiden, The Netherlands except where stated otherwise.
Koninklijke Brill NV incorporates the imprints Brill, Brill Nijhoff, Brill Hotei, Brill Schöningh, Brill Fink, Brill mentis, Vandenhoeck & Ruprecht, Böhlau Verlag and V&R Unipress.
All rights reserved. No part of this publication may be reproduced, translated, stored in a retrieval system, or transmitted in any form or by any means, electronic, mechanical, photocopying, recording or otherwise, without prior written permission from the publisher. Requests for re-use and/or translations must be addressed to Koninklijke Brill NV via brill.com or copyright.com.

This book is printed on acid-free paper and produced in a sustainable manner.

Contents

Figures

Notes on Contributors

Carma Casulá

is a visual artist and free-lance photographer. PhD in Fine Arts at the Universidad Complutense de Madrid (UCM). Specialized in Photography at IED Milan and ICP New York. Her artistic projects focus on anthropization and rootedness. Has exhibited her photos + installations in Spain, France, Italy, Germany, Denmark, Russia, USA and Costa Rica, among other places. Combines R & D in *Art, Ecology and Empathy* with teaching Photography at university level.

Anne Douglas

is a Professor Emerita from Gray's School of Art, Aberdeen where she led a programme of doctoral/postdoctoral research for artists and producers into the changing nature of art in public life. She has published extensively on key themes including improvisation, artistic leadership, the place of art in ecology and more recently Dada in contemporary art.

Bárbara Fluxá Álvarez-Miranda

(www.barbarafluxa.com) is a multidisciplinary artist, PhD in Fine Arts at the Universidad Complutense de Madrid (UCM), and researcher and teacher at the School of Fine Arts of the UCM. She is currently a research member of the R + D + I Project: Interactions of art in the technosphere of Ministerio de Ciencia, Innovación y Universidades - HAR2017-86608-P (2017–21).

Carmen Flys-Junquera

has recently retired as an Associate Professor of American Literature and Ecocriticism from the University of Alcalá, Madrid, Spain. She was the founder now member of the research group in ecocriticism GIECO (Franklin Institute, Universidad de Alcalá) and head researcher of the funded grant underlying this publication. She founded and still is Editor in Chief of the journal, *Ecozon@*.

Chris Fremantle

is a researcher and producer. He lectures at Gray's School of Art, Aberdeen and at Edinburgh College of Art. He established ecoartscotland in 2010 creating a new node in the international network of art and ecology practitioners. Projects he has been producer for, including the Harrisons' *Greenhouse Britain: Leading Ground, Gaining Wisdom* (2007–09), have won multiple awards.

Juan Carlos Galeano
is a poet, translator and essayist born in the Amazon region of Colombia. Extensive fieldwork on symbolic oral narratives in the Amazon basin resulted in his collection *Folktales of the Amazon* (2008) and the documentary films *The Trees Have a Mother* (2008) and *El Río* (2018). His poetry, including *Amazonia* (2003, 2012) and *Yakumama and other Mythical Beings* (2014), has been anthologized and published in international journals. He teaches Latin American poetry and Amazonian Cultures at Florida State University.

Lorraine Kerslake
holds a PhD in children's literature and ecocriticism and teaches at Alicante University, Spain. She is author of *The Voice of Nature in Ted Hughes's Writing for Children* (Routledge, 2018) and has published widely on children's literature and ecocriticism. She is an active member of the Spanish research group in ecocriticism GIECO (Franklin Institute, Universidad de Alcalá) and leading researcher of the project "Angels of the ecosystem?" (GV/2020/029) which focuses on the works of female writers who represent nature and non-human animals in literary texts of fiction.

Beatriz Lindo Mañas
is a PhD student in the American Studies program at the Franklin Institute, UAH. She is a member of the research group in ecocriticism GIECO (Franklin Institute, Universidad de Alcalá) and collaborates as an editorial assistant in the journal *Ecozon@*.

Lucía Loren Atienza
holds a Bachelor in Fine Arts from the Complutense University of Madrid. Professor of Fine Arts at Nebrija University. Her line of research revolves around contemporary art practices that integrate a socio-environmental reflection of our relationship with the territory, articulating collaborative praxis, social intervention processes and integrating these experiences in inclusive educational contexts.

José Manuel Marrero Henríquez
is a poet, writer, essayist, and senior lecturer in comparative literature and literary theory at the Universidad de Las Palmas de Gran Canaria (Spain) and member of the research group in ecocriticism GIECO (Franklin Institute, Universidad de Alcalá).

Imelda Martín-Junquera
is Associate Professor at the Department of Modern Languages and currently its Chair at Universidad de León, Spain. Her fields of research and interest are Chicano and Native American Literature and Culture from a posthuman perspective. She is a member of the research group in ecocriticism GIECO (Franklin Institute, Universidad de Alcalá) and main editor of the journal *Estudios Humanísticos: Filología*.

Elena Sánchez-Vizcaíno
is the Performing Arts Coordinator at TAI Centro Universitario de las Artes. She is and has been a professor in Arts Administration both in Spain and the US. Her research focuses on audience reception, social integration, and community building. Her most recent publication is "Using Thermography to Study Audience Engagement during Theatre Performances" in *AJAM* May 2020.

Irene Sanz-Alonso
is an Assistant Professor at the Department of Modern Languages of the University of Alcalá. After finishing her PhD in 2014, her research has focused on science fiction, fantasy and ecofeminism. She is a member of the Spanish research group in ecocriticism GIECO (Franklin Institute, Universidad de Alcalá) and is the secretary of the journal *Ecozon@*.

Scott Slovic
is University Distinguished Professor of Environmental Humanities at the University of Idaho, USA. He served as founding president of the Association for the Study of Literature and Environment (ASLE) from 1992 to 1995. In 2020, he completed a twenty-five-year term as Editor-in-Chief of *ISLE: Interdisciplinary Studies in Literature and Environment*. His twenty-nine monographs and edited books include, most recently, the co-edited collection *Reading Cats and Dogs: Companion Animals in World Literature*.

Diana Villanueva-Romero
is Associate Professor (*Profesora Contratada Doctora*) in the English Department of the Universidad de Extremadura (Spain) and a member of the research group in ecocriticism GIECO (Franklin Institute, Universidad de Alcalá) as well as of the group "Lenguas y culturas en la Europa moderna: discurso e identidad" (CILEM; Universidad de Extremadura).

INTRODUCTION

Imaginative Ecologies: Inspiring Change through the Humanities

Diana Villanueva-Romero, Lorraine Kerslake and Carmen Flys-Junquera

> Perhaps that old pair of antagonists, science and poetry, can be persuaded to lie down together and be generative after all (107).
>
> WILLIAM RUECKERT

•••

> Everything is connected to everything else (35).
>
> BARRY COMMONER

This book[1] has been a long-term project, one that has undergone many transformations since its origins, which date back to 2015 when a group of Spanish researchers from various fields—environmental ethics, arts and ecology, ecocriticism and literature—joined forces thanks to funding granted through a research scheme of the Spanish Ministry of Economy and Competitiveness. The title of this project, *Environmental Humanities. Strategies for Ecological Empathy and the Transition Towards Sustainable Societies*,[2] tried to capture what was then a bold attempt in many ways, particularly in Spain. Not only did this project aspire to introduce a fairly new discipline in Spain, but it did so from a highly interdisciplinary stand by gathering a collection of perspec-

1 This work was partly supported by Proyectos I+D+I Grupos de Investigación Emergentes, Conselleria de Innovación, Universidades, Ciencia y Sociedad Digital, Generalitat Valenciana. Grant Number GV /2020/029.

2 "Environmental Humanities. Strategies for Ecological Empathy and the Transition towards Sustainable Societies"; Subproject 2: "Stories for Change" HAR2015-67472-C2-2-R (MINECO/FEDER).

© DIANA VILLANUEVA-ROMERO, ET AL., 2022 | DOI:10.1163/9789004501270_002
This is an open access chapter distributed under the terms of the CC BY-NC 4.0 license.

tives from environmental ethics, arts, and ecocriticism and by paying attention to the need to understand the contributions of visual and literary cultures in promoting social change through the activation of empathy towards the environment using imagination as a tool. Such perspective acquired momentum after the celebration of the first International Conference on Environmental Humanities held at the University of Alcalá in 2018 under the title *Stories, Myths, and Arts to Envision a Change* in which a variety of practitioners from these different disciplines collaborated in what was an effort to create a space for disciplinary conversation and exchange.

The inspiring premise of these initiatives was that artistic works are themselves like cultural ecosystems,[3] creating awareness and interrelating ideas, visual and verbal linkages mixed with thought, values and traditions and lifestyles. In this book we take on the role of artistic works—visual cultures and literature—that promote environmental awareness through the exercise of the imagination and the analysis of these imaginative ecologies. Thus, 'imaginative ecologies' can be defined as ecologies of the mind moved by the affective connection between the writer/artist and the natureculture s/he is engaged with during the creative process of acknowledging the call of the more-than-human in an anthropocenic world where boundaries are built between the 'us' and the 'other.' Imaginative ecologies can pave the way for creative encounters of affective knowledge between these polarities so as to make possible the type of reconciliation needed in order to overcome the perils brought about by the nonsensical detachment embedded in today's turbo-capitalism.

The "texts" represented in this book all create, in one manner or another, spaces of empathy and awareness, and foster more sustainable attitudes and practices, hoping to kindle emotion. In this sense, we are inspired by the affective turn which has taken place in the humanities and the social sciences since the mid-1990's (Athanasiou, Hantzaroula, and Yannakopoulos 5) and firmly believe in the need for literary and artistic work to be informed by "embodiment and environmental situatedness" (Weik von Mossner 6) if it wants to be generative of change. That is why this book tries to show that the response to the ecological crisis expressed in the multiplicity of phenomena that haunt

3 In "Art and the Ecosystem," Patrick Kangas also uses the word 'ecosystem' in relation with art, but his approach is focused on analyzing the artistic form of ecological models used by ecologists to represent ecosystems. Equally interesting is Pedro Alves da Veiga's formulation of "a new media art ecosystem model" in reference to his analysis of the ways in which different agents and their relationships influence the creation of a framework that can either facilitate or hinder the success of artistic projects. Here, however, the word 'ecosystem' is used in a metaphorical sense, as a way to suggest that the creative imagination is made of different forces that interact among themselves and with the outside world in producing a work of art.

the future of planet Earth—global warming, water pollution, deforestation, biodiversity loss, etc.—does not necessarily have to come exclusively from the sciences, which always seem to opt for a detached, rational comprehension of the pressing questions of today's world. It can also arise from collaboration between arts and sciences, a methodology that, as it has already been acknowledged (Kueffer, Lässer, and Hall 2017; Kueffer et al. 2018), characterizes what has come to be known as environmental humanities. For this reason, this book gives voice to the practice of intellectuals, writers and artists from different nationalities who see themselves as participants of a global conversation on the role literature and the arts can play as facilitators of the type of social and cultural change needed in what Steffen, Crutzen and McNeill (2005) refer to as Stage 3 of the Anthropocene, a stage in which we can either continue doing business as usual or become the solution (619).

The truth is that it is not enough to simply understand the workings, for instance, of a complex process like climate change, for its rational apprehension does not mean we will commit ourselves to stopping it. What reaches our minds must necessarily move our hearts at some point in order to impel us towards the consecution of significant, long-lasting change. The scientific model is based on observation, description, testing and hypothesizing among other things. Writers and artists are experts in bringing us to our senses because they work through the muscle of empathy. They also observe and describe, test the reception of their work by their audience, and even hypothesize when they dare to create worlds of fantasy to make us see ourselves in another way, reconfigure who we are and what we represent in this world. Writers and artists are also scientists but of a subtler kind. Both groups are philosophers of the human soul trying to establish in their doings a relationship with the more-than-human world in which we all are interdependent.

This affinity between the sciences and the arts as two interlocked epistemologies is further stressed by tracing the history of the term "scientist." Stefan Collini indicates in a reprint of C. P. Snow's famous text on the cultural division into arts and sciences that "the term scientist was first proposed in 1834 along the lines of artists" (qtd. in Roberts 5). He explains that: "The lack of a single term to describe 'students of the knowledge of the material world' had bothered meetings of the British Association for the Advancement of Science in the early 1830s, at one of which 'some ingenious gentleman proposed that, by analogy with artist, they might form scientist.'" (qtd. in Roberts 5).

The contact zone between arts and science resonates in the conversations contained in this book as the expression of a long-lasting dialogue of the inquiring human mind with nature. Perhaps our brains need more than objective knowledge in order to formulate alternatives and understand the marrow of the problems that affect us every day. Interestingly, environmental activism often finds a loyal ally in writers and artists all over the world. Such alliance is underlined by Donna Haraway in her work on the Chthulucene where she highlights the effectiveness of science art projects like the Crochet Coral Reef project[4] to become "a case of multispecies becoming-with, a model system in which scientists, artists, ordinary members of communities, and nonhuman beings become enfolded in each other's projects, in each other's lives" (M35).

From this co-working of disciplines, it could be inferred that the meticulous and methodical work of scientists who dissect problems in order to efficiently formulate solutions is not enough. Literature and the arts in their multiple, peculiar enactments create *imaginative ecologies* in the sense that they translate into imaginings "the interrelationship of organisms and their environments."[5] The use of the word ecologies in the title of this volume serves to emphasize the webs of the relationships implicit in the different ways of knowing and experiencing the natural world. These ecologies are sustained by the human imagination, that potent generator of possibilities.

Thus, the term 'imaginative ecologies' is an expression that refers to the products of our imagination, those imaginings produced in our interaction with the non-human other. They are understood as the object of exploration of environmental humanists—historians, literary scholars, anthropologists, sociologists, philosophers, etc.—who perceive the need to explore the human imagination in its multifarious expression to recognize how humans negotiate their belonging to systems of naturecultures. As Brian Boyd points out, as social beings that we are, the arts appeal to our species preferences, regardless of the modifications by each local culture. Stories, he claims, "foster our ability to […] imagine possible actions in the face of any eventuality" (loc. 4672). Moreover, as cognitive scientist Lisa Zunshine argues, stories provide us with knowledge and understanding as well as creating "new forms of meaning for everyday existence" (164). Thus, we as humans tend to make sense of the world and relate to it through stories. Therefore, our relationship with the natural world can be transformed thanks to the stories we create about it. Or, by applying the artistic lens, might we not be able to bring ourselves to seeing what was before unknown to us or concealed from our sight?

---

4 https://crochetcoralreef.org

5 *The Merriam Webster Dictionary* defines 'Ecology' among other things as "a branch of science concerned with the interrelationship of organisms and their environments" (*Merriam Webster Dictionary*. Accessed 10 July 2020). Its words are borrowed to illustrate the concept behind the expression *imaginative ecologies*.

Several authors have already analysed the power of cultural artefacts, works of art and literature, to produce imaginative ecosystems where the human mind working through its empathic powers achieves a better, more profound understanding of itself in relation with the more-than-human world. William Rueckert distills his vision in his seminal article "Literature and Ecology: An Experiment in Ecocriticism" where he "[applies] ecological concepts to the reading, teaching, and writing about literature" in order to produce what he refers to as "literary ecology" (107). For him poems as well as other artwork are the "equivalent of fossil fuel (stored energy)" (Rueckert 108). Thus, any cultural artefact referred to in this book—a short story, a painting, a sculpture, an installation—can be understood in the same way. In line with Rueckert's tenets, the German ecocritic Hubert Zapf proposes "a theory of imaginative literature based on the paradigm of cultural ecology" (Loc. 129) which is a field of transdisciplinary studies where "ecology and culture, biological life science and the sciences of mind, systems theory and textual theory, natural and cultural evolution" are brought together (Loc 146). Like Rueckert, he is aware of the fact that other art forms besides literature "could equally be included in such a cultural-ecological analysis" (Zapf Loc 235) for all of them share "artistic potency" (Zapf Loc 284). That is, literature and the arts "[are] not only *ergon* but *energia*, a living force field of transformative energies which both subvert and enhance human consciousness and existence in rationalized modern societies" (Zapf Loc 306).

Thus, these imaginative discourses, these landscapes of imagination that writers and artists produce, can help us generate more sustainable forms of co-habitation. Perhaps we can learn from these creative laborers the way to disentangle our potential for fantasizing, for putting our imagination in practice. Not in vain Ursula K. Le Guin turns a passage of her book *Dancing at the Edge of the World* into a sheer defense of the powers of the human imagination since as she argues:

> Only the imagination can get us out of the bind of the eternal present, inventing or hypothesizing or pretending or discovering a way that reason can then follow into the infinity of options, a clue through the labyrinths of choice, a golden string, the story, leading us to the freedom that is properly human, the freedom open to those whose minds can accept unreality. (45)

When this book first began its journey, nobody foresaw a global pandemic that would transform our society and the world as we knew it. Over the past months our entire life has changed shaping a future that seems anything but bright. The crisis that we are facing is part of a much broader problem, one deeply connected to our dysfunctional relationship with nature. Now, more than ever, imagination is key in envisioning change and rethinking the way we interact with the world around us. Despite the fact that the chapters of this book were all written and reviewed long before the pandemic, many of them address the need for our civilization to change its anthropocentric view, the necessity for a shift of perception and values which requires new narratives to inspire change. In retrospective, they read as a framework that can provide us with new approaches to nurture the web of life and rethink our relations, reconciling nature and culture, science and art and help us envision a holistic view of the world through environmental imagination.

Environmental humanities has a key role in imagining how society will change. As a transdisciplinary approach it constitutes a valuable tool of ecological awareness and ethical transformation. These chapters offer initiatives to both help understand the cultural roots of our environmental crisis and, in fostering the attitudes and skills necessary, help change the perception of humanity and its place on earth, and urge for a biocentric rethinking of human relationships to non-humans and reawaken our senses.

The book has been divided into three major sections, each addressing different aspects of our imaginative process, from different understandings of the context and the need for perceiving the problems, to the academic analysis and interpretation of creative works to the actual creative process of those imaginative discourses. In the process, we have dealt with the different discourses of various disciplines, literary studies, pedagogy, philosophy and the visual arts, yet beyond the different methodologies and perceptions, the objective is the same—how the different "texts" converge into an ecosystem striving for increased awareness and methods of reading affect and inspiring change. The book is daring in its form as it presents an imaginative mix of different academic types of inquiry, an ecology of methods such as interviews, research essays, critical and descriptive and a short story. Such multiplicity of approaches, we believe, is necessary to grasp environmental issues from a humanities-driven perspective.

Thus, the first part, *Humanists in Conversation*, comprises three different articles, which can be read as diverse conversations all of which communicate the need to understand the Anthropocene. They challenge us to rethink our own relationship whilst drawing on the importance of education as a key tool to communicate and understand environmental issues.

Scott Slovic's article "Environmental Humanities and the Public Intellectual" opens this section and places environmental humanities in relation to the importance of teaching. Up until now scholars in environmental humanities have tended to focus on small problems rather than larger and more disturbing issues. Slovic argues, very convincingly, that a shift in action is necessary and that academics need to take on a more public role in ecological literary criticism and call attention to more pressing issues. Both personal and political actions must be engaged. As Slovic urges, it is time, now more than ever, for an "absolute mobilization of our hearts and minds" and he challenges us to think differently. Drawing on the special Winter 2014 issue of *ISLE: Interdisciplinary Studies in Literature and Environment*, devoted to global warming and climate change, he offers various pedagogical strategies which can be applied to current environmental issues in order to put this into practice in the classroom and engage students (through academic writing and public testimony). As Slovic shows, for environmental humanities to be effective, the practice of environmental humanities scholarship and teaching must be viewed in the context of academic responsibility and accountability.

This is followed by Diana Villanueva-Romero's interview with Christof Mauch, co-director of the Rachel Carson Center for Environment and Society, in Munich. Since its foundation in 2009, the Rachel Carson Center has contributed to the development of environmental humanities and stands as one of the most prestigious institutions in the field. The conversation looks at its history and development to explore how the centre has become a pioneering model. The institution facilitates research in the field of environmental humanities and holds a series of workshops and public conferences as well as hosting a doctoral program "Environment and Society" in collaboration with the LMU Munich. As Mauch explains, key to the ethos behind the centre was the idea of bringing scholars with a common interest in the environment from different disciplines together. Through the conversation Mauch explores the difference between the terms "environmental humanities" and "environmental history" and discusses the future of environmental humanities in Europe as well as the relationship of humanities as a field with natural sciences. Like Slovic, Mauch also advocates for the importance of scholars from environmental humanities to work with natural scientists.

This first section closes with Juan Carlos Galeano's interview "El Buen Vivir is Harmony. A Conversation with Rafael Chanchari Pizuri." Galeano, who is better known for his poetry inspired by Amazonian cosmologies, looks with Rafael Chanchari Pizuri, an Amazonian philosopher and shaman, at the sophisticated narratives of indigenous Peruvian Amazon tribes and their relationship with globalization and climate change. The interview considers current environmental challenges whilst relating them to the complex knowledge systems and narratives that the Amazonians have developed from their ancestors to interpret globalization and climate change. The conversation analyses the notion of "El Buen Vivir," a concept which englobes the harmony between good living and happiness inviting us to reconsider our environmental crisis from a new perspective whilst looking at human relationships with nature, the river and other ecosystems.

Part two, entitled *Interpreting Eco-visions*, comprises articles which read as imaginative discourses to understand the Anthropocene and offer an academic analysis and interpretation of creative works. The first article "Environmental Imagination and Wonder in Beatrix Potter" by Lorraine Kerslake seeks inspiration in Rachel Carson and Lawrence Buell, drawing on the reader's ability to experience a sense of connection with the environment and looks at the role that children's literature can play in imagining a more sustainable society and the importance of instilling a sense of wonder to the natural world in children. The chapter explains the close and necessary relationship between science and literature in the life and work of Beatrix Potter and how that close tie brings about a more realistic, non-sentimentalized awareness of nature for children. Through Potter's imagination and art, literature and drawings, children are both engaged with but also made aware of the not always peaceful cohabitation between species in the ecosystem.

This is followed by Chris Fremantle and Anne Douglas's article "Foregrounding Ecosystems: Thinking with the Work of Helen Mayer Harrison and Newton Harrison" which draws on *The Serpentine Lattice* (1993) and literary and visual devices such as ethnopoetics, presenting us with an alternative way of understanding the world and our existence, fundamental for the wellbeing of the web of life and ourselves. The chapter highlights the complexity of the Harrisons's artwork, combining visual arts, poetry, economics, ecology and politics and how this artistic imagination can shock audiences into striking realization that in turn can effect change.

Irene Sanz-Alonso's chapter "New Worlds Beyond Reality: Imagined Futures in Laura Gallego's *Las Hijas de Tara*" explores the values of alternative worlds and the societies portrayed in literary genres such as fantasy and science fiction as an alternative example of cohabitation. Sanz focuses on well-known Laura Gallego's novel *Las hijas de Tara* [Tara's Daughters], another work aimed at young adults. The writer uses her imaginative and pedagogical capacity to illustrate the negative consequences of an extreme nature/culture dualism. She creates an alternative futuristic world where nature and technology are at loggerheads with each other: both worlds are in continuous war against each other and the inhabitants of each are completely ignorant of the other. The journey of two girls from opposing worlds leads them to dispel their prejudices, understand and help each other, pointing to a lesson of tolerance and more balanced attitudes.

This is followed by Imelda Martín's chapter, "Simon Ortiz's Narratives and Joy Harjo's Poems: Towards Regenerative Societies and New Worlds" which uses the works of Ortiz and Harjo as windows which open to new possibilities and focus on the recovery of a "storytelling" tradition, representing an escape towards a transition of creating regenerative communities and new worlds. This chapter highlights how the stories and poems by Ortiz and Harjo draw new maps and paths in various directions in order to regenerate a harmonious balance and dialogic interspecies ethics. The tales shift from traditional Native American perspectives to the recreation of more sustainable and respectful attitudes which gather the old together with the new.

The section closes with Barbara Fluxá's piece "When Matter Takes a Position: Post-anthropocentric Landscapes in Contemporary Art," which advocates contemporary art as a transdisciplinary approach and offers a new cultural paradigm as an alternative to our systemic global crisis and the displacement of humanity. Fluxá illustrates how the artistic imagination, coupled with technology and scientific models, creates or recreates geological, natural and technological processes which have or could alter this Earth. She focuses on numerous artists and installations which in different manners recreate geological processes. The recreations clearly affect the viewers who, at times, unwittingly become performers by interacting physically with these installations, and are propelled into an awareness of these processes and their implications. The result goes beyond the simple representation of nature to look at artists in collaboration with those who work in the study of the solid earth and soil sciences to propose post-anthropocentric landscapes.

The last part, *Inspiring Change*, pays special attention to the creative process of imaginative discourses revolving around the idea that cultivating an awareness of the land may help to construct communities that, with time, will interact with their surroundings in a more sustainable way. It opens with artist Carma Casulá's contribution, "Sense of Place as an Enhancer of Empathy and Ecological Consciousness in the Baix Llobregat" whose site-specific work looks at landscape and the emergence of an ecological conscience among the inhabitants of the Baix Llobregat, an ambivalent space where conservation of spaces of significant natural, touristic and agricultural tradition co-exist with environmental degradation derived from intense industrial exploitation. As a photographer, Casulá's project is aimed at making this situation visible by creating a photographic series which she refers to as "forming a Memory Bank of Farmers in Spain." The collection is organized around two goals: first, offering a Physical Map of the area, and second, creating an Affection and a Connectivity Map based on the results of her field work with the local people through which she tried to understand the maze of relations that underlies their attachment to the place where they have traditionally lived and work. Hence, relationality becomes the organizing trope of a chapter where the author also voices her criticism of contemporary systems of living marked by a "disproportionate disaffection" and impels the reader to "[activate] biocentric attitudes."

This is followed by Carmen Flys and Beatriz Lindo's interview with the earth-based spiritual writer and activist Starhawk, aptly entitled "Building Stories to Change the World." The interview draws on Starhawk's novels and looks at the role of using narratives to envision change and reinvent new worlds through stories. This conversation shares with the previous one an attention to community, although of an experimental kind given the impact of Starhawk's novel *The Fifth Sacred Thing* (1993) where the artist-activist explores new ways of being in the world by describing what it would be like to live in a place ruled by norms inspired by "usefulness, beauty, sustainability, nourishing for the earth (or not damaging) and nourishing for the soul." Starhawk emphasizes the importance cultural and social tools have in addressing pressing issues of environmental importance such as climate change. She defends the power of imagination and ritual to instil change and offers a hopeful perspective on the potentiality of humans to regenerate themselves into better coparticipants of planet Earth.

In line with this message, Elena Sánchez-Vizcaino and Lucía Loren's article "Eco-interactions: Art and Community" offers insight to answering the question of how the arts can contribute to creating a more sustainable society. In doing so, Elena Sánchez-Vizcaino analyses Lucía Loren's work and her experience and interaction with environmental art through local communities. Loren's interventions, which she likes to refer to as "interactions," always invite local people to contribute with their knowledge and sensitivity to the place. Such is the case of *Al hilo del paisaje / Landscape Thread* (2007) done in collaboration with the people of Santa Lucía de Ocón, La Rioja (Spain). This is so because Loren, as Sánchez-Vizcaino explains, conceives her work as the result of a collaboration with the inhabitants of the place where her art will be integrated in its landscape. In her participatory model of art creation, Loren finds a potential to provoke an emotional reaction as well as a consciousness-raising. This gives her projects the potential to "[communicate] what is not visible to reason" and in so doing to serve as an inspiration for the future.

Last, but not least, the book closes with an epilogue in which José Manuel Marrero-Henríquez offers in "Chickens like Celebrities" a short story which takes place during a transatlantic flight when the protagonist, whilst contemplating a lunch menu, suddenly decides to become vegetarian for the well-being of the planet. The short story itself is an act of the author's imagination but it also deals with the imagination of the protagonist and how this inspires a personal decision to change. Marrero-Henríquez's short story meaningfully embodies the role that imagination can play in producing new responses to daily activities, such as eating meat, which may represent a more thoughtful appraisal of the role to play by humans in the world.

In the fields of literature and the arts, the importance of imagination to engage the reader and viewer and contribute towards change is more pertinent now than ever. Changing our view of humans' place in the world could actually save our planet. Imagination is perhaps the most important tool we have to do so. Not only can it engage us, but it can also make us aware of the impact of our actions and thus help to change our attitudes. The following chapters, from a variety of fields inherent to environmental humanities, therefore share an inherent conviction that imagination and the humanities can play a significant role in heightening environmental awareness and inspiring changing attitudes to promote transformation. As editors we hope these creative discourses may provide readers with ideas to inspire resilience towards the global environmental crisis that we are now fully immersed in. It is never too late to commit oneself to rethinking our relationship with nature and aspire to the type of harmony one of the contributors to this book, Rafael Chanchari Pizuri, describes from the last remnants of the Amazonia. Perhaps one day, in the not-too-distant future, the notion of "el buen vivir" will become a reality. Until that day comes, we will have to make do with finding hope and seeking ways to imagine change such as those suggested in this book.

Works Cited

Alves da Veiga, Pedro. "Towards a New Media Art Ecosystem Model." September 2005. PDF file. *ResearchGate*, doi:10.13140/RG.2.1.3551.5369. Accessed 16 April 2021.

Athanasiou, Athena, Pothiti Hantzaroula, and Kostas Yannakopoulos. "Towards a New Epistemology: The 'Affective Turn'." *Historein* 8 (2008): 5–16.

Boyd, Brian. *On the Origin of Stories: Evolution, Cognition and Fiction*. Belknap Press, 2009. Kindle edition.

Commoner, Barry. *The Closing Circle: Nature, Man, and Technology*. 1971. Courier Dover Publications, 2020.

"Ecologies." *Merriam-Webster Online Dictionary*. https://www.merriam-webster.com/dictionary/ecology. Accessed 20 July 2020.

Haraway, Donna J. "Symbiogenesis, Sympoiesis, and Art Science Activisms for Staying with the Trouble." *Arts of Living on a Damaged Planet: Ghosts and Monsters of the Anthropocene*, edited by Anna L. Tsing et al., U of Minnesota P, 2017, pp. M25-M50.

Kangas, Patrick. "Art and the Ecosystem." *The American Biology Teacher*, vol. 60, no.1, 1998, pp. 20–26.

Kueffer, Christoph, et al. "Applying the Environmental Humanities." *GAIA-Ecological Perspectives for Science and Society*, vol. 27, no. 2, 2018, pp. 254–56.

Kueffer, Christoph, Katharina Thelen Lässer, and Marcus Hall. "Applying the Environmental Humanities: Ten Steps for Action and Implementation." *GAIA*, 2017, pp. 254–56.

Le Guin, Ursula K. *Dancing at the Edge of the World: Thoughts on Words, Women, Places*. Open Road+ Grove/Atlantic, 2017.

Roberts, Adam. *The History of Science Fiction*. Palgrave Macmillan, 2016.

Rueckert, William. "Literature and Ecology: An Experiment in Ecocriticism." 1978. *The Ecocriticism Reader*, edited by Harold Fromm and Cheryl Glotfelty, U of Georgia P, 1996. 105–23.

Steffen, Will, Paul J. Crutzen, and John R. McNeill. "The Anthropocene: Are Humans Now Overwhelming the Great Forces of Nature?" *AMBIO: A Journal of the Human Environment*, vol.36, no.8, 2007, pp. 614–21.

Weik von Mossner, Alexa. *Affective Ecologies: Empathy, Emotion, and Environmental Narrative*. Ohio State UP, 2017.

Wertheim, Margaret, and Christine Wertheim. *Crochet Coral Reef*. The Institute for Figuring. https://crochetcoralreef.org Accessed 20 July 2020.

Zapf, Hubert. *Literature as Cultural Ecology: Sustainable Texts*. Bloomsbury, 2016. Kindle edition.

Zunshine, Lisa. *Why We Read Fiction: Theory of Mind and the Novel*. Ohio State UP, 2006. Kindle edition.

PART 1

Humanists in Conversation

CHAPTER 1

Environmental Humanities and the Public Intellectual

Scott Slovic

Abstract

This article places the discipline of environmental humanities in relation to Edward Said's "The Public Role of Writers and Intellectuals" (2002) and Marybeth Gasman's collection *Academics Going Public: How to Write and Speak Beyond Academe* (2016), arguing that the practice of environmental humanities scholarship and teaching must be viewed in the context of academic responsibility and accountability. During this age of global climate change, mass extinction, and various other environmental and humanitarian crises, it is particularly crucial for scholars in the environmental humanities to appreciate and act upon the societal relevance of our work, calling attention to important issues, tracing their historical roots, analysing communication practices associated with conditions that often defy easy description and narrative representation (too slow, too vast, too far away, too invisible), and probing the moral and spiritual dimensions of phenomena that seem to be primarily physical and political. In this article, I emphasize the examples of the Winter 2014 special issue of *ISLE: Interdisciplinary Studies in Literature and Environment* devoted to climate change and various pedagogical strategies for guiding university students and citizens to apply their voices (through academic writing and public testimony) to the urgent issues of the day.

Keywords

Environmental humanities, public intellectual, testimony, climate change, going public.

1 Silver Linings and the Unexpected Flowering of the Environmental Humanities

The entire discipline of ecological literary studies—and more broadly the environmental humanities—has flowered since the 1970s in response to

© KONINKLIJKE BRILL NV, LEIDEN, 2022 | DOI:10.1163/9789004501270_003

widespread recognition of the ecological crisis wrought by modern civilization. However, scholars in the field tend to isolate themselves for political or psychological reasons, focusing on small problems and topics that sidestep the vast and deeply disturbing issues of the day (climate change, military saber-rattling and actual conflict, the global refugee crisis, endangered species and the slow violence of extinction, global poverty, and many, many others). There are psychological reasons for avoiding such topics, especially if one is a humanities scholar—the blizzard of information we receive daily through the media can be overwhelming, the tools of writing and teaching may seem inadequate to the task of solving ecological problems or recalibrating societal actions, and the smaller, more immediate problems we face in our lives may simply seem more pressing.

I argue in this paper that two fundamental shifts need to occur in how environmental humanities scholars view our work as we try to lift it to a new level of cultural significance, both of which come from the idea of the "public intellectual": 1) Recognize on a visceral level that our discipline is a form of crisis response, not simply a timeless intellectual and aesthetic exercise; and 2) Cease writing merely for "the choir" and begin "going public" with what we know—and in turn inculcate this public-oriented attitude to up-and-coming scholars, including our students. In this article, I will place the practice of environmental humanities scholarship and teaching in the context of scholarly responsibility and accountability, drawing upon such sources as Edward W. Said's essay "The Public Role of Writers and Intellectuals" (from Helen Small's 2002 collection *The Public Intellectual*), the 2016 volume *Academics Going Public: How to Write and Speak Beyond Academe* edited by Marybeth Gasman, and my book *Going Away to Think: Engagement, Retreat, and Ecocritical Responsibility* (2008).

In her introduction to *Academics Going Public*, Gasman immediately argues that at least within the American context, the public intellectual has long been an accepted, indeed an *expected*, role for scholars and artists to play. She begins her discussion by citing Ralph Waldo Emerson's notion, in the 1837 essay "The American Scholar," that "A scholar's most important work is that of action" (Gasman 1). She does acknowledge, though, that this is a somewhat controversial notion in certain circles, with opinions ranging from the belief that public engagement is incumbent upon the majority of intellectuals (the reason for the protection of tenure that university-based academics often enjoy) to the idea that only rare intellectuals can and should rise to the level of "public intellectual" (1). I have argued in *Going Away to Think* that the tension between "saving" and "savouring" (between social engagement and contemplative retreat) lies at the very heart of ecocritical scholarship, at least in the case of my own work. I wrote:

> Academics are reputed to stand in campus lecture halls, dryly explaining topics of timeless significance. "Relevance" is a dirty word, a trivializing inclination. But in ecocriticism and related fields relevance is the Holy Grail--an intensifier of meaning, a trigger of passion. The work presented in this volume comes from my conviction that ecocritical responsibility requires both social engagement and reflective retreat. (xv)

Other ecocritics such as Lawrence Buell and Michael P. Cohen have suggested comparable ideas. Buell writes in *The Future of Environmental Criticism*: "Criticism worth of the name arises from commitments deeper than professionalism. Environmental criticism, even when constrained by academic protocols, is usually energised by environmental concern. Often it is openly polemical" (97). In his essay "Blues in the Green: Ecocriticism under Critique," Cohen argues: "by definition, ecological literary criticism must be engaged. It wants to know but also wants to do. Ecocriticism needs to inform personal and political actions" (27). To some extent, this tension between activism and reflection, at least in the American context, is an inheritance from one of the founding figures in American environmental writing: Henry David Thoreau. This is reflected in the very contrast between the author's profound meditations on the how the mind grapples with essential ideas about human experience and the natural world in *Walden* (1854) and engagement with injustice in "Resistance to Civil Government" (1849). Essayist E.B. White neatly captures the "two powerful and opposing drives" in Thoreau's work and in the eventual field of the environmental humanities as "the desire to enjoy the world ... and the urge to set the world straight" (238).

In the twenty-first century, the opportunities—and perhaps also the responsibilities—of the artistic and scholarly intellectual to take on public roles have increased. Certainly the ubiquity of instantaneous, affordable, and easy-to-use electronic media has made communication to various and vast audiences much easier than before. Edward W. Said noted this in "The Public Role of Writers and Intellectuals" (2002) when he wrote that "All of us should ... operate today with some notion of very probably reaching much larger audiences than any we could have conceived of even a decade ago.... [I]t is in the very nature of writing today" (28). Along with the enhanced ease of reaching hitherto unimaginable audiences through vetted media in print and online, the possibility of using unvetted, unscreened social media, from twitter to blogging, has completely altered the academic and literary landscape, the marketplace of ideas. This brings new opportunities and responsibilities for scholars and artists working in the environmental humanities arena. I will discuss several examples below to show how this evolving communication environment has influenced some of my own thinking and actions in recent years. Although this is an emotionally challenging time in history to be doing socially engaged work in the environmental humanities, one of the silver linings of this era is that the general public and civic leaders are now looking for ways to communicate more effectively with each other and to think more deeply about vexing issues. The environmental humanities, as a discipline, has much to contribute, and we are ready to step forward and meet these public challenges.

2 *The Story of the Climate Change Issue of* ISLE: Interdisciplinary Studies in Literature and Environment

On August 16, 2013, philosopher and environmental essayist Kathleen Dean Moore sent me an email asking if I would be interested in "publishing a 'call to writers' or 'call to conscience' in *ISLE*," an editorial that might inspire and compel our colleagues in the field of literature and environment to "think hard about how they are investing their time and abilities at an urgent point in history" when global warming has become impossible to ignore. Kathy and her husband had recently experienced 34-degree (Celsius) temperatures at midnight during a visit to Alaska's Denali National Park, and this was one of the motivating factors in Kathy's message to me. I wrote back, suggesting that we not only publish a Call to Writers in *ISLE*, the central journal in the field of ecocriticism, but that we try, on very short notice, to organize a special issue of the journal, following the model of the book *Testimony*, which Stephen Trimble and Terry Tempest Williams compiled in one month during the summer of 1995 in order to call attention to the protection of wilderness in southern Utah.

By August 18th, Kathy and I already had a draft of the Call to Writers and were ready to begin circulating it among individuals and organizations, asking for submissions by the end of September. Terry Tempest Williams herself placed the CFP on Facebook. Throughout the month of September, we received a torrent of messages and manuscripts. We ended up selecting manuscripts (only one per author) that seemed to represent an array of responses to global warming, more or less matching the categories of expression laid out in the Call to Writers. Most of these explicitly engaged with global warming, although some were allegorical or otherwise oblique. The initial Call to Action crafted by Kathy, which became the Call for Submissions to the journal, suggested the following categories of communication: The Drum-Head Pamphlet, The Broken-Hearted Hallelujah, The Witness, The Narrative of the Moral Imagination, The Radical Imaginary, The Indictment, and The Apologia. We received submissions that seemed to fit into all of these categories. In addition to poems and short stories and literary essays, we received several examples of scholarly (ecocritical) articles that engaged with climate change directly, ranging from theoretical pieces on the psychology of climate change and academic responsibility in the context of the current climate crisis to textual studies of climate film and climate fiction.

In *On a Farther Shore: The Life and Legacy of Rachel Carson* (2013), William Souder describes the era in the late 1950s when the technologies of DDT and nuclear fission had come together to spur "a new age of environmentalism [in which] the species that most needed our help would be us." At just this moment, writes Souder, "The world was ready to know this. All it would take was someone who recognized what was happening and who was willing to say it out loud" (264). When Kathy approached me in 2013 and proposed that we call the attention of the community of environmental writers and critics to climate change as an urgent topic, it seemed to me that we were experiencing the kind of moment Souder describes, in the context of global climate change after decades of work by Bill McKibben, Elizabeth Kolbert, and many others. The world, we felt, was at last "ready to know" the meaning of global warming.

In a sense, the climate change issue of *ISLE* occurred as the result of an all-hands-on-deck process—recognition of the emergent issue of global climate change led two editors to call our international colleagues to lend their voices to this vital issue. We felt—and still feel—this is a time for absolute mobilisation of our hearts and minds and voices. But it's important, too, to step back to take stock of what's working and what's not in ecocritical and literary engagement with climate change. This is an important way of improving the efficiency of our efforts and those of our colleagues. And that's why I have to ask whether writing—or art more broadly—is what's needed now, if our goal is to capture the attention of the wider public and move our fellow citizens to care and act. I realise that, as writers and literary professionals, what we do is write and talk/write about writing. But we saw graphic evidence in the 2016 American presidential election that there are huge numbers of people in the United States (and likely in other societies as well) who may not be readers—or at least who surely do not read the kinds of work that we produce. It's difficult to call attention to a crisis when we know that the vast, vast majority of our intended audience will be insensitive to the medium in which we utter such warnings. I have always spoken about the ripple effect of teaching and scholarly work—and even literary publishing—but the pebble causing the ripple is now starting to look impossibly small to me and the pond impossibly large.

For years I've had educated but non-literary friends telling me that I need to find other media for communicating important ideas to the broader public—media such as television or various mass-audience internet modes, such as a website or a blog. I've always resisted such pressure, saying, "That's not what I do." I think many of my friends and colleagues must feel the same way—meanwhile, millions of people watch reality television and end up placing more credence in reality TV and its creators than in science and art combined when it comes to vital issues. I think what friends have been telling me, and what I've never wanted to hear, is that a tremendous number of people (including voters of all political stripes) will never read literature or spend time viewing or listening to high art, no matter how beautiful and eloquent such expressions may be. This is especially true when the art and literature express viewpoints that the audience finds to be ideologically suspect or even disgraceful. Isn't the same thing true of us? I certainly don't want to spend my time soaking up messages that I find vile and dangerous. And thus, we are where we are: at a point in history where we and our fellow citizens, to the extent that we seek any information at all beyond our own minds, turn to ideology-confirming narratives, to stories that reassure us that we have the right and proper vision of reality rather than challenging us to think differently.

I understand the importance of empirical science and I realise that we are living during a time of severe biospheric crisis, yet I don't think we can simply maintain business as usual, reaching towards ever greater heights of eloquence and hoping that people who don't read will somehow, magically, wake up and start reading anything at all, especially the ideologically challenging and even scary work that we and our colleagues produce. We can continue "writing for the choir" because even the choir needs to be informed and inspired (this was the message of the talk I gave at a recent symposium in Havana), but we shouldn't fool ourselves that this work is really making a difference unless we find ways of reaching out to the many citizens who are not paying attention, do not want to pay attention, and will never be convinced to pay attention until there is some material, unavoidable reason to do so (and given the "slowness" of the climate crisis, as Rob Nixon would say, most people will never quite understand that much of what's going awry in the world is linked to the unfolding climate catastrophe).

If it's time to stop engaging in business as usual, I think people like us need to conceive new ways of communicating, new notions of what "the arts" are. Perhaps this means meeting our desired audiences at the level of the audiences rather than expecting "the great unwashed" to begin reading journals, (intellectual) magazines, and books. So, I would ask, how can artists reinvent themselves and how can their voices be heard by people who don't necessarily appreciate the arts, who have short attention spans, and who may disapprove of the messages (the entire world views) of the artists? This is why I've been experimenting with website design and face-to-face environmental writing (and thinking) workshops with diverse groups of people, ranging from high school students to Native American elders. I have not yet descended to Twitter, although I do have a Twitter account that I've never used. It's time ("All hands-on deck!") for writers and scholars like us to break free of our mental straitjackets and re-calibrate our voices to reach the people who need (who *we* need) to hear us.

As I look over the seven categories of climate change expression that Kathleen Dean Moore invented in 2013 and that we used to structure the Winter 2014 climate change issue of *ISLE,* I continue to recognize a large span of climate change discourse where there is room for a diverse array of literary expression. However, for some reason, writers who care about and are attuned to climate change are clearly not reaching a significant demographic group in American society (and this may well be true in other countries as well)—perhaps this is because such people are not readers (not reading anything at all), perhaps because climate change communication occurs in a register that is inaudible or non-credible for certain groups, and perhaps this is because many of us (including climate activists) are impervious to information that does not confirm our world views.

This latter idea became particularly clear to me as I listened to people during the 2016 campaign—how easy it is to find micro-narratives, especially online, that confirm our current ideology, regardless of what that is. If we live in a post-truth era (a time when civic leaders can spout outright fictions and be believed and when data can be withheld with no public consequences), why would any reasonable person inflict adversarial information upon herself or himself? Why would anyone not already concerned about climate change care to read or listen to or watch anything addressing the issue of climate change?

In light of the concerns I've just mentioned, the genres that we used in the *ISLE* special issue seem appropriately expansive to encompass many of the styles of climate change expression that I, as a climate-attuned person, attend to, but perhaps not the veins of expression that would attract the attention of people who are antagonistic to the concept of climate change. (I do think we could also mention a few other genres such as "The Science Made Plain" and "Gallows Humour" that would be useful for a broader range of readers.) Perhaps one of the lessons of the current political divide in the United States is that many citizens in this country (though not a majority!) just don't want to hear about climate change (or social justice). So, the question for people like us may be how to connect with that demographic—what are the styles of expression and the appropriate media that would enable substantive, ideology-cracking communication to occur?

One answer is obvious, if unpalatable: the tweet. But what "twitter genres" would enable substantive communication? Perhaps it is necessary to contemplate this question in order to reach "the great unwashed," as journalist and environmental writer David Quammen once described his target audience to me. One genre might be "Lucid Climate Science in 140 Characters." Another might be "Climate Jokes" (climate change is no joking matter, but many people who don't want to hear about serious, hard to visualise environmental or social problems will at least pay attention to good jokes)—and attention is one way to start the conversation. Also, what about a genre called "Denial" or "Enlightened Denial" or "Climate Resistance"? The latter could encompass both resistance to climate change and resistance to the idea of climate change, and once such viewpoints begin to be published side by side, the speakers/writers might begin interacting with each other and penetrating ossified viewpoints.

When Kathy and I published the Call for Papers for the *ISLE* issue, we had an immediate response from many potential contributors. But all of the respondents, as I recall, already belonged to "the choir." Those who had different views of the topic or who simply denied the importance or urgency of the physical process of climate change did not see the CFP or did not bother to share their voices with us. It's my sense that this may be the predicament of cli-fi literature and film, too—the adults who pay attention to such narratives are already in the fold. It may be a different story with young adult writing in the cli-fi mode, such as the excerpt from Brenda Peterson's YA (young adult) cli-fi novel, *The Drowning World*, that we included in the *ISLE* issue. One branch of cli-fi (and climate writing in general) that could be particularly promising is that branch that reaches towards younger readers who have not yet fully entrenched themselves in climate partisanship.

A genre that's somewhat related to one of the categories in the *ISLE* issue (The Witness) is what I would call "Personal Testimony." This is an area that I've been focusing on with my student writers for a number of years, not only in the context of climate change but in many other contexts, both trivial and significant. I always include personal testimony writing in my environmental writing classes, which includes training people (college students, colleagues, and the general public) to write testimony for public meetings, letters to the editor, letters to public and corporate officials, and op-ed pieces. One of the good things about testimony as a genre is that it invites statements from people of any social or political stripe, and it is a genre that empowers speakers to assert their particular views on issues of concern—not all testimonies are lucid and reasonable, of course, but all have a kind of intrinsic validity because they express the true feelings of the speakers. One hopes that the mere process of penning or typing words will force at least some writers to think more systematically and logically about their arguments than would be the case if they did not bother to write at all.

Related to testimony writing, my father and I have expanded the work of our 2015 book *Numbers and Nerves: Information, Emotion, and Meaning in a World of Data* by producing a website (www.arithmeticofcompassion.org) that explains the psychology behind why people care about various issues or fail to care and then indicating various ways of taking action in response to humanitarian and environmental crises, including climate change. The "Take Action" section of the website includes a section on writing personal testimony and several Youtube examples of such testimony. I think one of the most powerful examples is Marshallese poet Kathy Jetnil-Kijiner's presentation to the 2014 United Nations Climate Leaders Summit. A student of mine at an April 2016 environmental writing workshop at the University of Guam is a faculty member at Kapiolani Community College in Honolulu, where Jetnil-Kijiner was once a student—and she told me she is now teaching testimony writing with her students in Honolulu and sharing Jetnil-Kijiner youtube video (a short speech followed by a poem) with her classes to excellent effect: https://www.youtube.com/watch?v=L4fdxXo4tnY

Climate change has been interwoven as a subject in my writing and teaching for many years, though often indirectly. As I indicated in my contribution to the 2016 book *Teaching Climate Change in the Humanities*, edited by Stephen Siperstein, Shane Hall, and Stephanie LeMenager, I often use climate-related texts (such as *The End of Nature* and *An Inconvenient Truth* or shorter readings in anthologies) in my classes without explicitly talking about the climate aspects of the works unless students bring up the issue—my focus is often on literary form or broader issues of activism, and I find it more pedagogically powerful if I avoid heavy-handedly pointing to the gloomy topic of climate catastrophe. I have also used a chapter from Elizabeth Kolbert's *The Sixth Extinction* in my MFA nonfiction workshop at Idaho, but I didn't highlight the fact that the nearly extinct frogs that she was writing about in the chapter were imperilled largely because of climate change—I focused on literary form and on Kolbert's ways of weaving science into narrative, and I waited for students to point out the climate change link in the selected chapter. My approach to teaching climate change in university classes is often indirect—I put relevant material in front of my students and wait for the students to recognise the connections between the text and what they understand, consciously or unconsciously, to be happening to the Earth.

I am still coming to terms with climate change vis-a-vis in my own itinerant lifestyle. I travel frequently to give lectures and teach at environmental humanities events throughout the world. Author Bill McKibben was accosted by right-wing stalkers during the 2016 political campaigns in the U.S. regarding the inconsistencies between his climate activism and his use of fossil fuels, etc.—he wrote about this experience in the *New York Times*. I wrestle internally with this same inconsistency, but I try to find ways to make my travel really count (as Bill said he does as well when he gave the opening talk at the ASLE conference in South Carolina about a decade ago)—giving numerous lectures at every destination and staying in touch with many people I meet, encouraging them through email in their environmentally conscious teaching and research. This constant, daily support for colleagues throughout the world is one of the ways I tax myself in order to justify my travel. Also, increasingly, I have been presenting lectures via Skype or by sending Quicktime video recordings that are played at conferences in lieu of my physical presence.

3 Teaching Testimony, Teaching Outreach

In September 1995, inspired by Wallace Stegner's "Wilderness Letter" (1960) to the incoming Kennedy Administration (in which Stegner, as a private citizen, used his voice to speak about the importance of wilderness as an idea in American culture), Utah writers Stephen Trimble and Terry Tempest Williams quickly assembled short pieces by twenty writers into the limited edition volume titled *Testimony: Writers of the West Speak On Behalf of Utah Wilderness*. Stegner's letter helped to articulate some of the core ideas of what became the Wilderness Act four years later. Thirty-five years after Stegner wrote his letter, Trimble and Williams printed 1,000 copies of *Testimony*, and these were hand delivered to each member of Congress and to members of the press. After a press conference at the United States Capitol on September 27, 1995, a journalist asked what Trimble and Williams would do if the book made no difference in prompting support for wilderness preservation in southern Utah. The authors/editors responded: "Writers never know the effect of their words. [...] We write as an act of faith" (second edition 7). In fact, when President Bill Clinton signed into law the act designating the Grand Staircase-Escalante National Monument on September 18, 1996, he displayed a copy of *Testimony*, stating, "This little book made a difference" (*When Women Were Birds* 157).

Testimony has made a difference in other, less obvious ways as well. It inspired an entire wave, a social movement, of testimony collections, many of these gathering not only the voices of accomplished writers and politicians, but the words of "ordinary citizens," including students and first-time writers. I joined the movement myself in the early 2000s when Ranger Robert Moore from Great Basin National Park in Nevada enlisted me to help her compile a book that we eventually titled *Wild Nevada: Testimonies on Behalf of the Desert* (2004) with the goal of supporting public discussion of the meaning and importance of arid roadless areas in the State of Nevada, where I was living at the time. Thirty-one people contributed to the book, ranging from Senators Harry Reid and Richard Bryan to Shoshone elder Corbin Harney. Similar volumes have sought to mobilise support for the Arctic National Wildlife Refuge in Alaska (*Arctic Refuge*) and Petroglyph National Monument in New Mexico (*Voices from a Sacred Place*).

As a university professor and itinerant lecturer and writing instructor, I have also taken the testimony movement on the road, so to speak. When I began teaching Environmental Writing in the University of Idaho's Semester in the Wild Program in the fall of 2013, I incorporated a special unit on engaged citizenship and testimony writing at the end of the class, reckoning that even if my undergraduate students (most of whom were natural resource sciences majors) might never decide to become professional writers, all of them would spend their lives as citizens—and I hoped they might choose to be *engaged* citizens, confident in their ideas and voices and willing to write letters and opinion pieces and, if appropriate occasions arose, to stand up and speak at public meetings. I coached my students to write short testimonies on subjects they really cared about and to identify the audiences they were writing for, ranging from university officials to parents, and sometimes to corporate leaders. These are the essential facets of the testimonies my students write (and present to our class):

- Short, personal statements (300-500 words);
- Begin with a brief self-introduction, emphasizing one's connection with the issue;
- Incorporate story (brief presentations of experience that bring places and specific issues to life);
- Describe landscapes, seascapes, encounters with individual phenomena or people;
- Briefly articulate arguments or concerns, combined (if possible) with story;
- Try not to be overly abstract—accentuate the local and the personal.

I often begin my workshops by introducing participants to the building blocks (description, narration, and reflection) of what I call "the personal essay of environmental experience." We read and discuss several examples of such essays, including Scott Russell Sanders's "Buckeye" (1995) and Annie Dillard's "Living Like Weasels" (1982). We also take a look at examples of testimony, such as the statement before the U.S. Senate subcommittee on the Utah Public Lands Management Act of 1995, that Terry Tempest Williams included in her book *Red: Passion and Patience in the Desert* (2001). I find that workshop participants quickly begin to internalise the value and the style of such language if they hear the words coming out of their own mouths. We take the time to read these essays aloud, moving around our group, paragraph by paragraph.

Over time, word of my Environmental Writing workshops, including the personal testimony assignment, began to spread among my colleagues throughout the United States and abroad. In recent years, in conjunction with my lecture trips to various parts of the world, I have taught a three-day workshop in the Malaysian jungle (under the auspices of Universiti Putra Malaysia) and two-hour workshops at Istanbul Kultur University in Turkey and International Islamic University in Islamabad, Pakistan. In 2017, either on my own or with colleagues, I taught nine of these brief workshops, some of them specifically for environmental scientists or other targeted audiences, such as Native American elders, and others for the general public. The essential goal of these writing workshops is to empower participants to express their own knowledge and concerns—and to contribute their voices to public discussions of important topics. After I taught a day-long Environmental Writing on the Beach pre-conference workshop at Tumon Bay before the April 2016 International Conference on Island Sustainbility, I was pleased to see that the University of Guam's Center for Island Sustainability put the poem about invasive rhino beetles that undergraduate Arielle Lowe wrote for the workshop up on its website, amplifying the power of Arielle's voice, as she expresses concern about global climate change by describing a local impact (the spread of destructive beetles) on her island.

Through our individual voices, our idiosyncratic personal experiences and viewpoints expressed in the form of story, we have the ability to convey what we most care about—our concerns, our attachments. We do this most effectively when we use the highly particularised language of story, often in tandem with more formal scientific or professional information and analysis. Australian environmental journalist William J. Lines has written eloquently, in a 2001 essay about the danger of monetizing the value of nature, that "People exploit what has a price or what they conclude to be merely of value." He continues: "[…] they defend what they love. Love cannot be priced. But to defend what we love we need a particularising language, for we love what we particularly know" (*Literature and the Environment* 372). Here Lines echoes the famous words of Wendell Berry, who wrote similarly in his essay "Word and Flesh" that "Love is never abstract. It does not adhere to the universe or the planet or the nation or the institution or the profession, but to the singular sparrows of the street, the lilies of the field [...]" (200). All too often when we write and speak about the natural world, we drift into jargon and data, into recycled code words, into the abstraction of numbers. Such discourse has its purposes, including efficiency and the ability to capture the broader context of specific cases. But language, the language of image and story, also has its place in our efforts to communicate about natural resources and our deep attachments to the more-than-human world. When we teach the language of public testimony to our fellow citizens, we empower them to participate more fully in public conversations about important subjects. Empowering the public in this way is a meaningful and appropriate role for scholars and writers to play.

In aggregate, the stories ordinary citizens tell have the potential not only to help communities cohere, to help them overcome the divisiveness that is all too common in countries like the United States today, but to influence large-scale policy. In other words, communities of average people throughout the world have the power to exert in pursuit of the common good. Daniel C. Taylor, Carl E. Taylor, and Jesse O. Taylor offer various examples of this in their book *Empowerment on an Unstable Planet: From Seeds of Human Energy to a Scale of Global Change* (2012): "Genuine human progress involves the use of human hands, hearts, and minds to do what we can with what we have, here, today. Every human being, even the most impoverished, has a portion of discretionary energy to direct. Begin with that, grow it" (xv). This vision is closely in sync with the philosophy of bioregionalism, the social movement that emerged in the American West in the 1970s, with a particular focus on the validity and viability of locally based resource management in specific watersheds. As Gary Snyder describes bioregionalism in a 1992 lecture (and subsequent article in the *San Francisco Examiner*), "Watershed consciousness and bioregionalism is not just environmentalism, not just a means toward resolution of social and economic problems, but a move toward resolving both nature and society with the practice of profound citizenship in both the natural and social worlds. If the ground can be our common ground, we can begin to talk to each other (human and nonhuman) once again" (235).

In addition to including personal testimony units in my Environmental Writing classes and teaching special workshops that focus specifically on testimony writing, I have found myself in recent years increasingly focusing on the concept of reframing scholarship for public audiences. This includes both using ideas from the environmental humanities and related disciplines in op-ed pieces for newspapers and magazines and training students and colleagues (often from the environmental sciences) to write press releases based on their fundamental research in ecology, natural resource management, and humanities concerns such as environmental values and aesthetics. In the spring of 2017, three fellow English professors and I taught a two-hour workshop titled "Environmental Science Writing for Public Impact" at the Center for Environmental Research, Education, and Outreach at Washington State University; attended by twenty-five graduate students, post-docs, and faculty members in the sciences, the workshop focused on preparing salient and accessible statements about important new research that might catch the attention of journalists, who would then be in a position to publish reports on this work for the general public. During this workshop, in addition to providing specific examples of powerful environmental writing and general principles of such writing, we used Kat Stein's chapter "How to Write an Influential Press Release" from Gasman's *Academics Going Public* in order to provide clear guidance regarding the elements of a good press release (including telling the *story* of the research and avoiding jargon and numerical data).

I have also begun incorporating this public dimension into my scholarly seminars in the humanities, asking my graduate students to write not only technical papers with specialized journal audiences in mind, but then to prepare press releases for journalists on the subject of their research about novelistic treatments of endangered species and biosemiotic approaches to climate change literature. I tell my students that if their work is significant enough to appear in a scholarly journal during this age of rampant social despair and environmental degradation, it should also be important enough to be of interest to general readers. If we are not doing work of social and environmental relevance, then we are, as the saying goes, playing chamber music as the Titanic sinks into the icy sea.

Works Cited

Berry, Wendell. "Word and Flesh." *What Are People For?* North Point, 1990, pp. 197–202.

Buell, Lawrence. *The Future of Environmental Criticism*. Blackwell, 2005.

Cohen, Michael P. "Blues in the Green: Ecocriticism under Critique." *Environmental History,* vol. 9, no. 1, 2004, pp. 9–36.

Dillard, Annie. "Living Like Weasels." *Literature and the Environment: A Reader on Nature and Culture*, edited by Lorraine Anderson, Scott Slovic, and John P. O'Grady. 2nd ed., Pearson, 2013, pp. 5–8.

Gasman, Marybeth, editor. *Academics Going Public: How to Write and Speak Beyond Academe*. Routledge, 2016.

Huser, Verne. *Voices from a Sacred Place: In Defense of Petroglyph National Monument*. Self-published. March 1998.

Jetnil-Kijiner, Kathy. "Statement and Poem by Kathy Jetnil-Kijiner, Climate Summit 2014 – Opening Ceremony" *YouTube*, 23 Sept. 2014, www.youtube.com/watch?v=L4fdxXo4tnY.

Kolbert, Elizabeth. *The Sixth Extinction: An Unnatural History*. Henry Holt, 2014.

Lines, William J. "Money." *Literature and the Environment: A Reader on Nature and Culture*. Pearson, 2013, pp. 367–73.

Lowe, Arielle. "UOG Student Arielle Recites Rhino Beetle Poem." *YouTube*, 9 May 2016, www.youtube.com/watch?v=V59XfiYuPXE.

McKibben, Bill. "Embarrassing Photos of Me, Thanks to My Right-Wing Stalkers." *New York Times*, 5 Aug. 2016. Online edition.

Moore, Roberta, and Scott Slovic, editors. *Wild Nevada: Testimonies on Behalf of the Desert*. U of Nevada P, 2004.

Said, Edward W. "The Public Role of Writers and Intellectuals." *The Public Intellectual*, edited by Helen Small, Blackwell, 2002, pp. 19–39.

Sanders, Scott Russell. "Buckeye." *Literature and the Environment: A Reader on Nature and Culture*, edited by Lorraine Anderson, Scott Slovic, and John P. O'Grady. 2nd ed., Pearson, 2013, pp. 186–91.

Slovic, Scott. *Going Away to Think: Engagement, Retreat, and Ecocritical Responsibility*. U of Nevada P, 2008.

Slovic, Scott. "The Elephant in the Room: Acknowledging Global Climate Change in Courses not Focused on Climate." *Teaching Climate Change in the Humanities*, edited by Stephen Siperstein, Shane Hall, and Stephanie LeMenager, Routledge, 2016, pp. 163–69.

Slovic, Scott, and Kathleen Dean Moore, editors. Global Warming Special Issue. *ISLE: Interdisciplinary Studies in Literature and Environment*, vol. 21, no. 1, 2014.

Snyder, Gary. "Coming into the Watershed." *A Place in Space: Ethics, Aesthetics, and Watersheds. New and Selected Prose*. Counterpoint, 1995, pp. 219-35.

Taylor, Daniel C., Carl E. Taylor, and Jesse O. Taylor. *Empowerment on an Unstable Planet: From Seeds of Human Energy to a Scale of Global Change*. Oxford UP, 2012.

Trimble, Stephen, and Terry Tempest Williams, editors. *Testimony: Writers of the West Speak on Behalf of Utah Wilderness*. 1st ed., West Wind Litho, 1995.

White, E.B. "A Slight Sound at Evening" (1954). *Essays of E.B. White*. Harper & Row, 1979, pp. 440–48.

Williams, Terry Tempest. "Statement Before the Senate Subcommittee on Forest & Public Lands Management Regarding the Utah Public Lands Management Act of 1995, Washington, D.C., July 13, 1995." *Red: Passion and Patience in the Desert*. Pantheon, 2001, pp. 72-78.

Williams, Terry Tempest. *When Women Were Birds: Fifty-four Variations on Voice*. Picador, 2012.

CHAPTER 2

Humanities in Transition in the European Context

Interview with Christof Mauch

Diana Villanueva-Romero

Abstract

This chapter is based on an interview conducted with Dr. Christof Mauch, Director of the Rachel Carson Center for Environment and Society in Munich, Germany during the summer of 2017. Its goal is to explore with him the contribution of this centre to the environmental humanities since its foundation in 2009 as well as his perspective on the development of the field from the privileged vantage point of managing one of the most prestigious institutions in the field worldwide and a pioneer in Europe. This piece intends to offer an overview on the origins and growth of the environmental humanities in the European context aided by my personal academic experience as a Spanish ecocritic as well as readings of state-of-the-art bibliography in the environmental humanities. With this, without forgetting to mention its many challenges, I eventually aim at giving the reader a sense of the diverse nature of the environmental humanities as well as its potential in a time when an urgent need to transitioning to new ways of living and doing academic work are needed.

Keywords

Environmental history, environmental humanities, environmental studies, interdisciplinarity, Rachel Carson Center (RCC).

The Rachel Carson Center (RCC) is currently the world's largest Centre for Advanced Study in the Environmental Humanities. It is part of Ludwig-Maximilians-Universität Munich, Germany's highest rated university. The Deutsches Museum in Munich is the RCC's local partner. The centre was established in 2009 with project funding from the Federal Government of Germany. The initiative to build this centre goes back to environmental historian

© KONINKLIJKE BRILL NV, LEIDEN, 2022 | DOI:10.1163/9789004501270_004

Christof Mauch. Diana Villanueva-Romero spent three months at the RCC in 2017. Towards the end of her tenure she interviewed Christof Mauch (CM) in Munich.[1]

The interview was conducted at the Rachel Carson Center, Munich, August 7th, 2017.

I: *Can you tell me about the origins of the Centre? How was it possible to create something like this? After two months here, I'm really impressed by what you have achieved; I see it as a model to follow, an example, so I'm curious to know how the centre started in 2009 and where the funding came from.*

CM: We were really lucky. There was a competition going on for about five years. The competition started in 2007 and during those five years 10 projects were selected, projects that would be supported, actually paid for, by the Federal Ministry for Research and Education. These 10 international centres were created to internationalize the humanities in Germany. The ministry felt that the humanities, more than the sciences, were too national in their focus. Moreover, representatives of small disciplines were encouraged to apply. I decided to apply with a proposal in the field of environmental history—a very small subject in Germany; even today there is no Chair in Environmental History anywhere in Germany. At the time I might as well have applied for a different topic, for example in the field of German-American relations, because my background was in transatlantic history. I had been running the German Historical Institute in Washington D.C. and I had lived in the U.S. for 14 years before coming to Munich. During my time at the German Historical Institute, environmental history became my hobby. I had organised about a dozen international conferences in the field of environmental history because I felt that this was really a fascinating topic. When I returned to Germany a couple of people approached me and asked whether I would start an institute or a project in the field of environmental history, and they pointed me to the ministry's competition. So I decided to apply under this call.

1 This interview was possible thanks to a Mobility grant by the Junta de Extremadura which allowed me to be a visiting scholar at the Rachel Carson Center for Environment and Society from June to August of 2017. I am also deeply grateful to Prof. Mauch for his generosity in sharing with me his time and ideas.

Fortunately, at a reception, I met the research director of the Deutsches Museum in Munich, Helmuth Trischler. I asked him whether he would want to be part of the proposal. Helmuth was excited. The combination of a university and a museum made our application unique and strong. In the end we were lucky. We applied and we got funding for six years and later on for an additional six, altogether 12 years.[2]

I: *Why the name Rachel Carson?*

CM: My hope was that we could give the centre a non-German name, and the name of a woman rather than a man. So many German foundations and institutes carried the name of a German male scholar: Leibniz Gemeinschaft, Humboldt Foundation, Goethe Institute, Böll Foundation, Adenauer Foundation, etc. Our goal was to think of somebody who would be known outside of Germany, an academic who had gone beyond the ivory tower. The name should signal that our centre was a research centre but also a centre with a political mission. In fact, at the Carson Center we are trying to reach out to the public through public symposia and lectures, through occasional exhibitions, and through our digital portal which today reaches hundreds of thousands of people around the globe.[3] The funding that we received from the ministry was meant for fellowships. But at the German Historical Institute in Washington, I had learned that outreach was an important complement to research. So, from the beginning the idea behind the Carson Center, in contrast to other projects that received funding from the ministry, was to strengthen the outreach component and also to play some role in current political debates.

---

2 For an adapted version of the proposal that led to the establishment of the RCC see Mauch, Christof, and Helmuth Trischler. "International Environmental History: Nature as a Cultural Challenge." *RCC Perspectives* 2010, no 1. doi.org/10.5282/rcc/5581. In a public talk at the opening of the Rachel Carson Center Christof Mauch explained what the RCC was all about: Mauch, Christof. "Notes from the Greenhouse: Making the Case for Environmental History." *RCC Perspectives* 2013, no. 6. doi.org/10.5282/rcc/5661.

3 An international essay competition that would demonstrate the topicality of Rachel Carson's legacy led to hundreds of responses from all over the world. The two winning essays were published in Culver, Lawrence, Christof Mauch, and Katie Ritson, editors. "Rachel Carson's *Silent Spring*: Encounters and Legacies." *RCC Perspectives* 2012, no 7. doi.org/10.5282/rcc/5597.

Moreover, we hoped to transcend borders. From the beginning we were interested in flows and transfers of people, of materials, of resources, of flora and fauna. After all, natural environments ignore national borders. In fact, our goal was to study and compare human-environment relations across space and through time. We felt it would be interesting to study how societies, cultures, experts in different parts of the world and in different periods in history had dealt with similar challenges. History tends to be very descriptive, but if you start to compare across time and space then you go almost automatically beyond description. Comparisons provide you with an opportunity for unique analysis and this provides you with the potential for lessons.

I: *What about other disciplines besides history?*

CM: Initially our institute had a focus on international environmental history. However, from the beginning we felt that it did not make sense to focus on history alone. In fact, some of the most boring conferences were the ones that brought together historians only. (Even worse were those that brought together German-speaking historians only). During phase I, from 2007–2013, we hosted about 50% historians and 50% non-historians. We found that the best way to come up with new and exciting research questions was by bringing people from different disciplines together: Scholars who have something in common—in our case an interest in the environment—but who also differed in their methodological approach. I remember, for instance, an informal discussion when one of our fellows, an ethnobotanist who happened to be the President of the Global Diversity Foundation, was challenged by an ecocritic from Canada who asked him about the value of the concept of diversity. This led to a heated debate among fellows from different disciplines. In the end we decided to organize a multidisciplinary workshop on the topic "Why we value diversity" which was published in our *Perspectives* series. We realized that multidisciplinary settings make us come up with fundamental questions that we would never ask in a single-discipline setting. Many of our workshops were both multidisciplinary and cutting-edge. We were, I think, the first who did a workshop on neurohistory. The workshop started in a hospital with students under a scanner. They were presented with different images of urban and rural landscapes, land and sea, etc. We asked scientists to help us explain what happened on a neurological level. As humanities scholars we normally study texts and artifacts, but we ignore what happens in our brain as a result of environmental stimulation. In our workshop and the resulting *Perspectives* publication, edited by Edmund Russell, we tried to understand how we can integrate neurological understandings into our humanities research. Last year, to give another example, we discussed the concept of human niche construction in a workshop with philosophers, architects, social scientists and others. Niche construction is a well-known concept in biology. The standard example are beavers who create dams and reconstruct and reshape their environment—often dramatically—according to their needs. Humans are doing something similar, on many levels, including for instance the air-conditioning of our living spaces.[4] Our workshops have had an enormous potential in bringing people together from different disciplines and from different countries.

I: *How did the Centre evolve over time?*

CM: In the beginning we received our funding mainly for what the ministry called "free space." Scholars were given a stipend so that they could sit at the desk and write books. That was the basic idea. We introduced weekly lectures, our lunchtime lecture series in which our fellows introduced their work to their peers and to the interested public. After some time we established the format of works-in-progress, weekly meetings that gave the fellows an opportunity to discuss their work in written form through precirculated papers. In order to reach a broader public, we worked on our social media, on digital exhibitions, and finally also on non-virtual exhibitions in cooperation with the Deutsches Museum. We began to publish our own journal, *Perspectives*, and an English and a German language academic book series.[5] We also launched "Green visions" which featured monthly viewings followed by discussions with experts. In the meantime, fellows and staff had

---

4 For more information on the concept of human niche see Ertsen, Maurits W., Christof Mauch, and Edmund Russell, editors. "Molding the Planet: Human Niche Construction at Work." *RCC Perspectives: Transformations in Environment and Society* 2016, no. 5. doi.org/10.5282/rcc/7723.

5 *Perspectives: Transformations in Environment and Society* (online journal), http://www.environmentandsociety.org/perspectives (Series editors Christof Mauch, Katie Ritson, and Helmuth Trischler), since 2019 *Environment in History - International Perspectives*, Berghahn Books, New York/Oxford (Series editors, Christof Mauch, Dolly Jørgensen, David Moon, and Helmuth Trischler), since 2009; Umwelt und Gesellschaft, Vandenhoeck und Ruprecht, Göttingen (Series editors Christof Mauch and Helmuth Trischler), since 2009.

developed their own formats, in particular discussions on such topics as gender and the environment, on risk society, on water issues etc. So, within three or four years we had totally expanded and reconstructed the original focus on "free space."

I: *Did the organizational and the thematic focus change over time?*

CM: During phase I of our project we offered six research foci. They included transformation of landscapes, ecological imperialism, environmental knowledge, resource exploitation, cultures of catastrophe and risk, and environmental ethics and politics.

Setting this up at LMU was anything but easy. Our university had no programme in environmental studies. The Carson Center was something like a *creatio ex nihilo*. If you had come to Munich, let's say ten years ago, and you had told anybody that Munich would set up an environmental centre, it would have sounded like science fiction. I'm a professor of American cultural history, I had organised conferences in history when the RCC started. I had published a few articles and edited a few books in this field, but nobody would have thought of me or my co-director (Helmuth Trischler is a historian of technology) as leading figures in environmental history. We started from next to nothing. But there was a belief that questions that are related to the environment are of real importance for the future of humankind, questions like: How are we using our resources? How will we feed an ever-growing population? Etc. For questions such as these, history can serve as a great compass.

Now, in order to make our project sustainable beyond the six plus six-year funding phase from the Federal Ministry, we proposed a doctoral programme and a master's programme. We ended up creating a truly international doctoral programme with students from every continent, but we could not establish a master's programme in environmental studies.[6] This proved to be too difficult within a university with a very weak central structure and about 50,000 students. Instead we created a certificate programme that gives master students from currently more than 40 disciplines a second "degree," a certificate in environmental studies.

In order to get funding for a second phase we had to come up with new ideas and programmes. Among other things we added new types of fellowships—Outreach Fellowships (for filmmakers and writers for instance); as well as interdisciplinary fellowships for scholars who come from different disciplines but work together on a common project. We encouraged the creation of an

6 In 2019 the RCC received funding from the Volkswagen Foundation to create a master programme, MOOCs, workshops, visiting professorships and exhibitions through the Volkswagen Foundations funding line "University of the Future."

alumni organization, the Society of Fellows, and we now have fellowships that are jointly sponsored by the alumni and by the RCC. The idea of the alumni fellowships is that former fellows come back and "give back"—for instance by organizing workshops, teaching a class or doing a virtual exhibition in their field of expertise.

Another *novum* were the regular discussions with environmental experts—policy makers, members of NGOs, environmental publishers, the heads of natural history museums and zoos and national parks, the City Planning Office etc. We also decided to include the students in outreach projects. One project was Ecopolis—a big student exhibition about the environmental history of Munich to which citizens and city planners contributed. This exhibition (part of it is available online) explains, for instance, why beer gardens have chestnut trees.[7] It reveals that Munich's Olympic Park sits on a mountain of World War II rubble, and it explains that the City of Munich buried a whole village under a mountain of garbage. The exhibition also elucidates how the rise of the Alps has shaped the terrain of the city of Munich, and what it took to re-naturalize the Isar river.

Also, in order to be sustainable beyond the ministry funding phase, we recently established a foundation for which we hope to gather tax-deductible donations for some of our projects. And we focus more and more on local research projects, including one that will start in 2018 about the National Bavarian Forest, Germany's first National Park.[8]

Other regional activities include place-based workshops that bring students and fellows from different disciplines together: cultural historians, hydrologists, ecologists and geologists have camped out in places south of Munich and explored human-nature relations as part of a multi-day seminar with our certificate and doctoral students. This year we also did a place-based workshop along the Danube, in collaboration with our partners from the University of Wisconsin in Madison. We invited 18 experts from many different

7 http://www.environmentandsociety.org/exhibitions/ecopolis-muenchen2017. In 2019 the Ecopolis project was continued, with a second exhibition titled: Ecopolis München 2019: Environmental Stories of Discovery. The project received LMU's Teaching Award. Parts of the exhibition are available online and so is the catalogue Laura Kuen Gesa Lüdecke, and Christof Mauch, eds. *Ecopolis München: Umwelt- und Entdeckungsgeschichten. Eine Ausstellung des Rachel Carson Centers for Environment and Society*. Munich: Rachel Carson Center for Environment and Society, 2019. http://www.environmentandsociety.org/exhibitions/ecopolis-muenchen-2019.

8 The project resulted in seminars, field trips and a multidisciplinary publication on the 50th anniversary of Germany's first national park. Marco Heurich and Christof Mauch, *Urwald der Bayern: Natur, Geschichte und Politik im Nationalpark Bayerischer Wald*, Göttingen: Vandenhoeck und Ruprecht, 2020.

disciplines—engineers and architects as well as foresters and archaeologists—who helped us to explain the relations between the Danube River and its people in the past and into the future. Because our environmental study programmes bring students from very different disciplines together—humanities as well as social and natural sciences—students learn to speak about their own discipline in a way that is accessible. They are not only talking to their own in-group but learn to find a way to communicate so that people with a very different background can understand them as well. Our 130 certificate students and 30 doctoral students are a small group but what they are doing is unique. Their community is truly international and interdisciplinary. No comparable programme exists anywhere in Europe today.

I: *Do you have partners outside of Munich?*

CM: It is important for us to have partners in other places. We work together with the Centre for Ecological History at Renmin University in China, for instance. We organize annual workshops there. We have a partnership with KAJAK, an environmental centre in Tallinn, Estonia. We organize mini-workshops with the University of Vienna. The University of Zürich and the RCC have a joint fellowship programme. We established a doctoral exchange with Kansas, we are about to set up a partnership with New York University in Abu Dhabi where one of our alumni is teaching. Currently, we are getting proposals almost on a weekly basis from scholars in different parts of the world who are hoping to collaborate with us or apply for grants together.

I: *So far, throughout this conversation, you have often used the expression environmental history, not environmental humanities, and I would like to know the reason for that choice. What is your perception of the way in which environmental humanities, as a field, is growing and also do you think European institutions can play a role in its development?*

CM: These are at least three questions and I think they are really important ones. First, you have addressed the point of nomenclature. When we started the Carson Center, it was officially called "Internationales Kolleg, Internationale Umweltgeschichte: Natur als Kulturelle Herausforderung." That seemed a bit long. Now "Rachel Carson Center" was potentially misleading, because, without a subtitle, it could have referred to a centre that deals with the work of Rachel Carson. There is already a Rachel Carson Institute in the U.S., so we needed a clarification. Also, Rachel Carson was a biologist, and we were doing environmental history and related disciplines. So, for a short time we called our centre the "Rachel Carson Center for Environmental Studies." Nobody knows that, but we actually had letterhead that read: "Rachel Carson Center for Environmental Studies." At the very time, when we were about to print the letterhead, there were two fellows at the centre who discussed the term "environmental studies." One of them was from Britain and the other one from Canada. One of them said, "Environmental studies means, for me, something scientific," and the other one said, "Environmental studies means, for me, something literary." Because of these potentially confusing connotations, we got rid of our brandnew letterhead design.[9] We decided to call the RCC "Center for Environment and Society." I have to say I liked the broadness of the term and the suggestion that our work was not merely scholarly but potentially political as well. In contrast, I never quite liked the term "environmental humanities." The reason why I don't like it—but I may very well change my mind—is because until recently almost everybody who calls themselves an environmental humanities scholar is a literary scholar. The Carson Center started out as a centre for environmental history. This is what our grant was for. And the term "environmental humanities" de-centres history. I realize that so many people now call their centres Environmental Humanities centres. Currently I still think we should keep the "Environment and Society" label, but it is also important to be pragmatic. Recently, we established a doctoral degree in "environmental humanities." As the Carson Center is part of LMU's literature and language faculty, I could not suggest a degree titled "environmental history" or "environment and society." Instead, I asked for permission to establish an environmental humanities PhD, and it was granted. Students who will receive such a degree can have a background in many different disciplines—anthropology, geography, literature, history. If you want to work outside of academia, in the environmental "field," it

9 Since 2017 the RCC has increasingly used the term "environmental humanities" for its projects and events. In July 2018 Christof Mauch convened the first ever International Summit in Environmental Humanities in Schloss Hohenkammer/Munich. The event brought together scholars from every continent, including a number of doctoral students and 26 directors of Environmental Humanities centers.

may be an advantage to be an environmental humanities scholar rather than, for instance, a scholar of modern history or theology. Depending on the future of the term "environmental humanities," we might decide that we change the name of the Carson Center and call it an Environmental Humanities Centre. Currently, however, I am afraid that the term does not do what we want it to do. The term is not inclusive enough. Many geographers who see themselves as scientists would not easily identify with "environmental humanities."

I: *And the question about Europe and its role?*

CM: Europe is special. I realised this when I was president of the European Society for Environmental History (ESEH). Members of our society at the time were publishing in twenty-four languages. I am always amazed how an organization like the EU can survive. It holds so many different countries together under one umbrella. Interestingly, in Europe, in contrast to many other continents, the natural sciences feature more prominently in environmental history conferences than elsewhere. Historical climatologists attend the European Environmental History Conferences and give papers. In America, many people work on cultural aspects and they discuss issues of national identity. In Europe in contrast there is no common cultural history. If you look at the programme of the most recent ESEH conference there are many different disciplines represented: geographers, anthropologists, political scientists, and also literary scholars. EASLCE, the European Association for the Study of Literature, Culture, and the Environment identifies itself much more with the environmental humanities than ESEH. But this may well change. I will admit that the term "environmental humanities" is not without some charm. There are aspects that I like about it. I like the fact that it brings together the physical world: the "environment" out there, alongside "humanity" and the "human world." When geologists or biologists talk about nature, they usually don't think of humans. So, the reference to the humanities is attractive because it includes both impact on humans and human perception. But the term "Environment and Society" suggests relevance beyond academia—perhaps more so than "Environmental Humanities." Environment and Society suggests outreach and policy orientation, and my hope is that our research will be relevant beyond the ivory tower.

I like the little essay that Frank Zelko wrote for a conference that we organised on "The Future of Environmental History" several years ago in Washington D.C.. Zelko says something like: Next time after you are done with your book, ask yourself whether there is anything that the non-specialist can draw from. Does history provide solutions that can be used in today's world? Can we write a relevant policy statement based on our study of the past? Zelko is a co-editor of a journal called *Solutions* and he urges us to think whether we might have solutions for current-day challenges. Humanities scholars, in general, and history scholars, in particular, are very good at identifying problems, but once we have identified a problem we tend to leave it at that. Likewise, if somebody comes up with a solution, we are very good at critiquing the solution. We are really good at explaining that there are no easy solutions. We are good at demonstrating how complex our world is. But I agree with Frank Zelko who would claim that we become more relevant as environmental historians or environmental humanities scholars if we realise we can also help to provide solutions to contemporary challenges. We may be criticised for it, but our work is concerned with human-nature relations and indeed with the future of the planet. It would be good if our research could give direction and orientation for political decision makers. I think as environmental scholars, we have an obligation to identify ways for moving forward in a manner that is less destructive to our environment. As environmental historians we have learned that humans in different times and in different parts of the world have found ways out of alarming situations. We know about cultural responses that can provide us with hope. If we learn about these ways through our research, I think we have an ethical obligation in our world to intervene. Now I am using a military term "intervene," "intervention," and I realise that this is a problematic term. But what I mean is that we should intervene in the sense of taking part in political debates if we have relevant knowledge. Sometimes we see a lot of alarmism about developments that humans have handled fairly well in history. In such a situation, we should raise our hands and say, "look we've handled such situations before." What looks like an aporia, is not quite as gloomy as one might think. So, I really like this idea of Frank Zelko that, at the end of writing a book, we should write a memo or an opinion piece for a newspaper and discuss how we can make our world a better world. A better world can mean a more hopeful world but, in some cases, it can also mean a more concerned, a more alert world.

I: *From what you say, I see that a focus on the environment can bring hope to the humanities. In Spain, we have the sense that the humanities are in a time of crisis. It is difficult to get funding for projects in the humanities, so perhaps by incorporating the environment in what we do, we have a possibility of telling society how meaningful the humanities can be. So, what do you envision as the future of environmental humanities in Europe?*

CM: The short answer is I agree with you 100 percent. We have seen a recent concern for the environment and a surge in the environmental humanities programmes. This in itself is a cause for hope. We have come a long way within half a century. The first time that anybody could get a university degree in environmental studies was in the 1950s in the U.S. It was a natural science degree. Today, people—far beyond the circle of specialists—realise that our environments are fragile and vulnerable. We are today in a better situation as environmental humanities scholars than we have ever been: we have a community of potential listeners. People who look for orientation. I see the environmental humanities as orientation sciences. They help us orient ourselves because we narrate stories in which humans, minds, intentions etc. have a place. In some of the stories, we are victims of environmental change, in others we are drivers of positive change. We live in a time in which people are eager to listen to environmental humanities researchers, in part because we talk precisely about the role of humans in the environment. However, I am also worried. I do not see that our funding agencies—and particularly the EU—realise the value of the humanities. Unless EU officials change their advertisements for academic projects dramatically, there is a danger that we will become ornamental and indeed irrelevant. On the one hand, the EU is paying lip service to the integration of the humanities in their calls for research projects. Some natural scientists and engineers integrate philosophers, historians, lawyers or some other humanities scholars in their projects. However, typically humanities scholars play a peripheral role in almost every large EU project. Including a humanities scholar is like ticking a box: putting a humanities scholar on the roster gives the project a better chance of winning. The Carson Center often receives inquiries from natural scientists or engineers who want to include us in their proposals, but most of the time they have no idea about what the Carson Center is doing. All they want is the formal involvement of a humanities institute. If one wants to get funding from the EU in today's world, one better be a scientist or engineer, not a humanities scholar.

Funding goes to those who collect data and evaluate data and to those who do applied research. Environmental historians have, of course, also collected historical data. If a city is flooded, we, as historians, can provide data. We have records that show, for instance that the water level of the Danube in 2013 was higher than at any other point since the Middle Ages, with the exception of 1501. This information—and the 1501 data—may be useful for engineers. But if we reduce ourselves to the collection of data, we lose what makes us who we are: critical humanities scholars.

What we need on the EU-level is nothing short of a revolution. Currently, the proposals and calls are written in a science and engineering language. We need calls that speak to the strengths of humanities scholars. There should be calls for research projects that come out of an understanding of what humanities scholars can provide. We are good, for example, at explaining complex transformations over time. We are good at answering questions about what happens to people, to livelihoods, to flora and fauna when humans intervene, for example through deforestation, the planting of soy, monocultures, etc. We are good at answering questions about cultural, social and political responses to catastrophes—in different parts of the globe and at different times. We are good at identifying unintended consequences. We are good at imagining better worlds and good at analysing past utopias and dystopias. For some of our projects we would seek scientific knowledge and support. Remote sensing, for instance, could be helpful when it comes to an analysis of landscape transformation, but our skills and the value of our work is not in doing applied research. The questions we are asking are "why" and "how" questions, and our answers can rarely be reduced to simple data. We would not simply provide numbers and correlate them. Recently some scientists reported: "Look, in certain areas of the globe where a lot of water is used, income is low in agriculture." This may well be true, and it makes for good headlines. As humanities scholars and as environmental historians in particular, we would be critical of such a statement. We would ask who uses the water, how the water is being used, why this specific correlation was investigated. We would also think of alternative questions that one might ask, and we would always include what one could call "context": cultural context, social context, political context, legal context. Not just data. Politicians want data and executive summaries, short abstracts, and oftentimes a legitimation for their agenda. Under the current call for EU proposals no large-scale project will ever be led by an environmental humanities scholar. And there are reasons for it.

We are an academic community without a strong lobby. And we are small in numbers. Most politicians do not know that the environmental humanities exist. So perhaps we should indeed work towards a revolution. I think that our world would be a much better place if every university had at least two or three environmental humanities scholars or environmental historians. Our world would be a different world; universities would look different. A stronger presence of our field would create a new culture of understanding that would also be visible to people who are decision-makers. We have a long way to go, but we should not give up. I truly believe that the contextual critical thinking that the humanities can provide is needed today more than ever before, particularly when it comes to human-nature relations. Humanities scholars are good at providing a deep understanding of what scientific data might mean. For example, what impact does the use of pesticides or a transformation to monoculture have? What is the impact not just in terms of productivity, but also regarding the minds and outlook of people, social realities, environmental justice, visions, etc. The humanities are better than the natural sciences at providing a comprehensive picture of our world.

I: *What should the humanities relationship with the natural sciences look like? And how does the Rachel Carson Center handle this relationship?*

CM: What scientists are doing is extremely important and it will be important that environmental humanities scholars work together with natural scientists. But I contend that it is easy to mock the natural sciences when medical advice about how much coffee and how much wine we should drink, and whether it is healthy or not to drink a certain amount per day has changed dozens of times during my lifetime alone. Scientific results have a very short lifespan. The stories that humanities scholars write typically have a much longer lifespan. After all we write against forgetting: we give voice to humans and to human-environment relationships that would not be remembered without our narratives.

What I feel we need is an academic culture in which students and scholars don't think in dichotomies and hierarchies between natural sciences and social sciences and humanities. We need people who can understand the importance of coming together and contributing to a larger project that reaches beyond disciplinary boundaries. We should try and come together first and foremost in order to contribute to a better, a just, a diverse, a sustainable future. These broad goals are often forgotten in our research agendas. The environmental studies programme of the Rachel Carson Center may be a small step in this direction. Our students have to attend courses in natural sciences, humanities and social sciences. They come from many different disciplines. The environmental studies programme forces our students to communicate across disciplinary boundaries. They are all enrolled in a regular master's programme, but they attend environmental classes as well. They get a second degree, a certificate. Because of their different backgrounds, they learn how to communicate with peers who come from very different disciplinary cultures. Our Certificate Programme in Environmental Studies will not save the planet, but if people would think in the way our environmental students think, then our world would likely become a better place. Altogether, I am optimistic. We are currently a small group, but we are a growing minority and we have never been in a better position to grow than today.

I: *I recently had an interesting experience applying for a Project Life, a European type of funding that was established to give funding to projects having to do with environmental issues. I was discouraged by scholars and by the Spanish authorities because my project was educational, so I can fully relate to what you are saying. We need a revolution. What do you think about this type of situation where a humanities scholar finds herself blocked by a long-established system of doing and understanding things?*

CM: I think we are victims of a neoliberal culture and a type of thinking in the European Union that gives priority to economic development. There are, of course, reasons for this. The last line of EU research calls came at a time of economic crisis. In order to legitimize research funding, it was necessary to come up with an economic reasoning. In the rationale of the EU, research needs to contribute to development and it needs to create profit. This type of thinking leaves little or no room for what you described as your own educational project. There is no direct monetary profit in educational research projects. If we criticize the EU, we should not forget, however, that the situation is not much better elsewhere. U.S. education is all about money. Universities are notoriously run like businesses. Students have to pay, alumni have to pay, and governments end up paying less and less. In a way it is understandable why the EU prioritizes applied research. The phenomenon is not EU specific. It is an expression of capitalist ideology and utilitarianism.

I: *Do you see a way out of the current situation?*

CM: I can see two ways out of the dilemma. First: a slow "march through the institutions." I think that there is a chance that over time EU

> bureaucrats will change their calls. I also believe—and I have seen this—that natural scientists, some of them at least, are fascinated by what the environmental humanities are doing. As humanities scholars we should get heavily involved, directly or indirectly in all the meetings that prepare future EU research calls.

The other option is to apply to non-EU funding agencies that are more open-minded. That way we might be able to, subtly and subversively, create awareness of the role of the environmental humanities and strengthen our role in public discussion. I would not underestimate the fact that our strength is in communicating science. In recent years environmental humanities scholars have very effectively gathered around the term Anthropocene. The momentum the term Anthropocene gained is fascinating. A group of geologists created the Anthropocene Working Group with the goal to formally establish a new geological epoch. That epoch is supposed to be located on the same hierarchical level as the Pleistocene and Holocene epochs, with the understanding that the Anthropocene terminates the Holocene. These geologists did not get any big grant for their project. They did not need it. They got free publicity for their project—way beyond anything that anybody could have ever imagined. The publicity came from hundreds of environmental humanities scholars around the globe, but also from artists, musicians, painters, sculptors, philosophers, historians, literary scholars, museums. The geologists got free advertisement for their concept from cities, from landscape projects, from people who work in conservation, because this term, a scientific term, grasped the imagination. It helped the Anthropocene Working Group to have humanities scholars and artists as their mouthpiece. Humanities scholars and artists have shown themselves to be the best analysts and interpreters of what the Anthropocene might be, and what human presence means for the future of the planet. One could go further and say that explaining the Anthropocene needs the humanities. This is a reason for optimism, and perhaps one day the EU will propose a call about the Anthropocene or similar concepts. That would give humanities scholars an opportunity to become project leaders.

I: *You have done some work on the Anthropocene at the Carson Center, haven't you?*

CM: Yes, the Carson Center developed and curated a Deutsches Museum exhibition titled "Welcome to the Anthropocene" that was visited by close to 300.000 people. The term Anthropocene is a broad term. It has topicality and it is future-oriented. We called our exhibition "Welcome to the Anthropocene." The title suggested that we are just now walking through the gateway of a new geological era. The future of the planet will be determined by what we as humans have done and will be doing in the future. The currency of the term Anthropocene has helped environmental humanities scholars get visibility. It has helped geologists as well. In recent years, geologists have often been criticized because of their inability to predict big disasters, such as tsunamis or earthquakes, and because of their involvement in environmentally destructive mining projects. The Anthropocene has given geology an altogether new identity, and some geologists have adopted and embraced that. I am claiming that we can learn from this phenomenon. It shows us that interdisciplinary and transdisciplinary projects can be attractive. It shows us that we have potential allies in politicians and decision makers and artists, and in natural scientists, in this case geologists, as well. We do not need armies of natural scientists as partners. But natural scientists who are open-minded are extremely valuable partners. The secret behind the success of the Swiss Network in Environmental Humanities which was established just a few years ago is that it includes a small number of highly motivated natural scientists. A few natural scientists is all it needs.

Currently, it looks like the environmental humanities have no ally on the EU level. But we are slowly gaining traction, in particular through university programmes. We are gaining traction through the ideas and activities of students who begin to think differently. New policies will certainly not come out of diplomacy alone. The Paris Climate Summit was important not because of its direct policy implications but because of its symbolism. Environmental humanities scholars are playing a major role in translating the goals of the summit into action. Perhaps some time in the future Brussels will understand that an EU project is not only valuable if it brings in money or leads to new patents, but also because it creates awareness and concern. When I say we need a revolution; I think it is basically going to be a grass roots revolution, but this revolution has already started—in all the environmental humanities centres that have sprung up like mushrooms over the last few years. Many of them are working on a shoe-string, but they have gained a lot of visibility, especially if they work with local and regional communities.

Works Cited

Gora, L. Sasha, editor. "Ecopolis München." Environment & Society Portal, *Virtual Exhibitions* 2017, no. 2. Rachel Carson Center for Environment and Society. doi .org/10.5282/rcc/7959.

Martin, Gary, Diana Mincyte, and Ursula Münster, editors. "Why Do We Value Diversity? Biocultural Diversity in a Global Context." *RCC Perspectives* 2012, no 9. doi .org/10.5282/rcc/5599.

Möllers, Nina, Christian Schwägerl, and Helmuth Trischler, editors. *Welcome to the Anthropocene: The Earth in Our Hands*, Munich: Deutsches Museum 2015.

Möllers, Nina. "Welcome to the Anthropocene: The Earth in Our Hands." Environment & Society Portal, *Virtual Exhibitions* 2014, no. 2. Rachel Carson Center for Environment and Society, www.environmentandsociety.org/node/6363.

Russell, Edmund, editor. "Environment, Culture, and the Brain: New Explorations in Neurohistory." *RCC Perspectives* 2012, no 6. doi.org/10.5282/rcc/5596.

Zelko, Frank. "Offering Solutions." "The Future of Environmental History: Needs and Opportunities," edited by Kimberly Coulter, and Christof Mauch. *RCC Perspectives* 2011, no 3. doi.org/10.5282/rcc/5623, 38–40.

CHAPTER 3

"El Buen Vivir" is Harmony with the Earth

Interview with Rafael Chanchari Pizuri

Juan Carlos Galeano
Translated by Vera Coleman

Abstract

Born in the Peruvian Amazon, Rafael Chanchari Pizuri is a philosopher and Amazonian from the Shawi ethnic group, whose spiritual ecological discourse, rooted in indigenous cosmovisions of the indigenous cultures of the Peruvian Amazon, foregrounds the current environmental challenges and complex symbolic narratives of indigenous Amazonians. He is also a shaman and a teacher, who has contributed a great deal to the education of indigenous bilingual teachers of many ethnic groups in the Zungarococha FORMABIAP-AIDESEP Educational Centre in Iquitos, Peru. This interview discusses the sophisticated knowledge systems and story cycles that Amazonians have developed over millennia to interpret globalization, climate change and human relationships with forests, rivers and ecosystems. It also considers indigenous notions of "el buen vivir" (good living) and education in Amazonia and the way these are represented in indigenous oral narratives and literary texts, which invite us to reimagine the world's environmental crisis from the perspective of Amazonian thought. The interview draws on multiple conversations over the past ten years, as well as interviews conducted for the films *The Trees Have a Mother* (2008) and *El Río* (2018), both directed by Juan Carlos Galeano.

Keywords

Amazonia, indigenous ecologies, cosmovisions, interculturality, *buen vivir*, Shawi.

I: *We are in the Amazon region and I cannot help thinking about the fundamental role that the land plays in indigenous cultures. This is why I would like to ask you to begin by sharing the cosmovision that the Shawi people have of forests and rivers.*

© KONINKLIJKE BRILL NV, LEIDEN, 2022 | DOI:10.1163/9789004501270_005

RAFAEL: The most important thing for us is the land where we were born and the rivers where we get our water. A Shawi child or adult can be happy where they are if the land provides them with everything. The earth is like a mother who gives you food and things to drink. Then people feel nourished, we feel comfortable.

As adults when we go around saying, "Oh, what a beautiful land this is, it has enough forests, enough fruits, it has medicinal plants." There are different types of land. For example, the kind with black soil is productive. Black soil is good for growing bananas, sugar cane, peanuts, beans, sweet potatoes, and squash. White sandy soil is good for manioc, pineapple and barbasco [*Lonchocarpus urucu*].[1] Black soil and sandy soil are what we want for our plants. On the banks of the rivers where we live there's black soil, and deeper down you find sandy soil and then clay soil.

We usually harvest only twice because the earth is like a person, it also gets tired. Two harvests and nothing more. We need almost every kind of soil. To give you an example, we use yellow clay soil for ceramics. Living like this, you feel that you lack nothing. Mother Earth gives you everything. Along with all of this, something fundamental for the earth is its veins, by which I mean its rivers. The earth and her water. That's why when our ancestors were looking for a place to live, they wanted it to be near the water, because rivers and streams are like the veins of the earth. Our ancestors were looking for good land and for the earth's veins.

I: *Since you mention the importance of the river and the water, I remember that in the documentary El Río that we filmed recently, you appear as a central figure, a philosopher of indigenous Amazonian cultures, who points out that water is an essential component. Could you talk to us more and elaborate on the importance of rivers and water in indigenous cultures? I would appreciate if you could tell us about the role of water in the spiritual life of the Shawi people.*

R: For me and for the Shawi child that I was, the river is very important. That's where I learned how to swim, where I learned how to fish, where I learned how to row. For Shawi children, the river is something wonderful. We can play in the river. We can develop many physical abilities, like going underwater and holding your breath, which is good for fishing later on.

---

1 Barbasco [*Lonchocarpus urucu*] is a species of plant containing poisonous chemicals that different indigenous groups use for fishing.

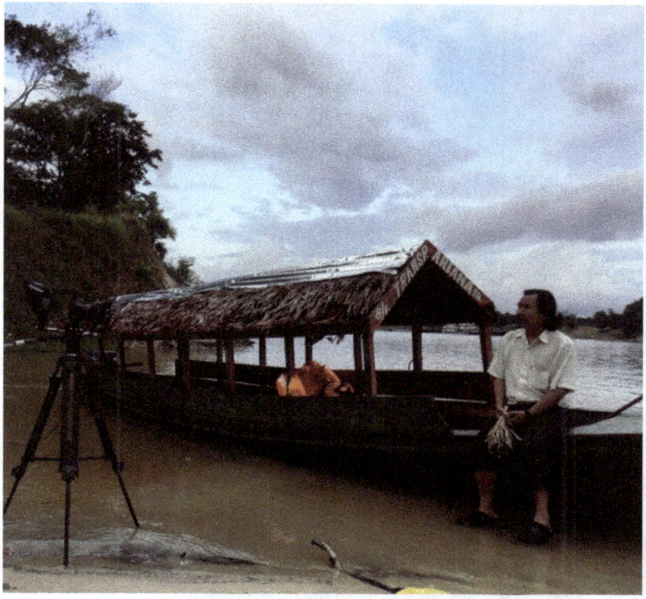

FIGURE 3.1 Rafael Chanchari Pizuri, Philosopher of the Shawi Nation at the Nanay River, during the filming of the documentary El Río, in the Peruvian Amazon, 2016. (Picture taken by Juan Carlos Galeano).

The water helps us a lot in our learning because we get to know the fish that live there and the birds that live and fish in the rivers. There we learn to handle canoes, manage rafts. It is a space for gaining knowledge. When you need to eat, you run to the river and get water for cooking. When you want to make masato [manioc beer], you get water from the river. The river in Shawi culture represents a source of life because many species live in the river. The river is where we get food and fish so our children grow healthy and strong. And when you want to move from one place to another, the river is your highway. In short, the river is life. You cannot be isolated from the river.

As I was telling you, the earth is like a human being with a circulatory system. It's the same. The river is a vein or an artery of the earth. As for the spiritual aspects that you asked me about, we are aware that spirits live in the river. There is the mother of the deep pools, the Yakuruna and the pink dolphins. There is also the Yakumama, which is the mother of water. You have to be careful with these spirits when you go to the river. We don't destroy the forest upstream and in the mountains, because that produces massive erosion, which disrupts the amount and flow of water in the pools where these mothers live, forcing them to leave the river and take the fish with them.

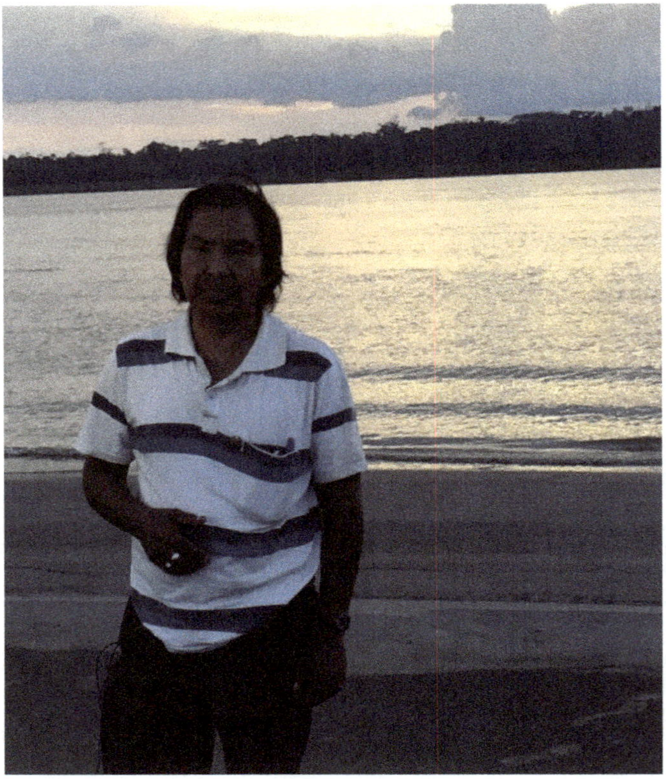

FIGURE 3.2 Rafael Chanchari Pizuri, Philosopher of the Shawi Nation at the Amazon River, during the filming of the documentary El Río, in the Peruvian Amazon, 2016. (Picture taken by Juan Carlos Galeano).

To mistreat the river is to hurt the spirits. And as for the spiritual side, it is thanks to the river that we can prepare our medicines. With medicines like tohé and ayahuasca, we can have a relationship with the spiritual beings of the water. They know, they already know you. Then you ask them, "I want to have wisdom to cure evil." These plants are not drugs, as some think. They are sacred plants that we have used for thousands of years and that give us the power to see the past, the present and the future. With these sacred plants, you can know what happens around you and in other places. Thanks to them, you can see the world as if you had it in your hand. For example, Shawi people go to the river beaches, they smoke their mapacho cigars, or take another plant, sleep and dream. The water spirit comes and gives them power to heal others. This is how we gain power from the river.

We also use tobacco to ask permission from the mother of the river so that we can go to a beach or go to a certain place to fish. These plants function as a means of communication. Thanks to these plants, we can navigate far from where we are. We can be here, but with the plants we become connected. With them we can see the relationships that exist among all the beings that populate the earth, including us. Nothing is separate in the world. Everything is connected. Human beings, the river, forests, the wind, the thunder.

The wind for us is a person, it's a spirit. These spirits have life. For scientists, perhaps, they are natural phenomena, but for us they are spiritual beings who hear you, understand you in different languages. The river comes into contact with the sun, the rain, with the wind, because they are part of the entire life system. The river is also a source of inspiration. We play with our imagination on the shores of the river, we learn to build a lot of imageries. We develop our imagination. Many thoughts and images spring from the river. We even imagine who created the river.

I: *In recent years, indigenous cultures have gone to great lengths to strengthen their identities, to survive and to persist culturally, wanting to ensure the life of people and the survival of species and places. I would like to ask you how the continuity and survival of your community happens in the context of our critical times, generated by climate change.*

R: As you know, the arrival of Europeans in the Amazon in the sixteenth century brought about a gradual and constant undermining of indigenous cultures. In the case of the Shawi people, who were conquered in those years, we still continue to fight above all to maintain our language, to maintain our life. Our culture thrives in a territory of vital importance to us. Because for us, the forests, rivers and animals are a source not only of sustenance, but also of knowledge and wisdom.

As I told you many years ago in the interview for the documentary *The Trees Have a Mother*, when the Spaniards came, they thought we were an inferior culture with no knowledge, no history and no wider significance. That we just lived to live and nothing more. Then the Spaniards said, "We need to civilise them. We need to put clothes on them. We need to give them machetes. We have to give them axes so they will work." In the old days, indigenous people didn't plant big chacras [plot gardens in the forest]. We ate from the forest. The Spaniards said, "These people are lazy, so we have to give them tools and teach them how to work." They didn't realise that we also had our knowledge.

If the Spaniards hadn't come, I don't know how indigenous people would be now. Perhaps we would have developed another type of knowledge. That's why the Shawi people are still trying to preserve their language, their culture, music and dance, their trees and medicinal plants. We want to continue speaking our language, but there's a trend towards castillianisation for indigenous peoples.

There's also the influence of other religions. There's influence of national and regional policies that try to change our identity. These situations influence us, creating in us a sort of multiple identity based on these influences and the way we articulate our cultural resilience and the existence of forests and rivers within that context. Because we want our social, political and religious identities to persist and develop, so that the negative policies of Western culture don't come to dominate and annihilate us. In that way we continue to persist inside Peru, so that we are not assimilated, we are not absorbed and made into mestizos.

I: *It seems to me that within such indigeneity there is a kind of mestizaje, an interracial or cross-cultural relationship.*

R: That's true. We can say that within the indigenous realm we are mestizos because we have mixed with the Awajun and Kokama ethnic groups and other members of indigenous cultures. But we make sure that our Shawi identity, our religious identity, is maintained and developed. On the other hand, the idea about the possibility of a Shawi government is also emerging now. But this idea hasn't gone deep enough, hasn't fully developed yet. But what would the autonomy of the Shawi look like in a country like Peru, one that is multilingual and multicultural? This remains to be seen.

Where are these trends heading? What are the difficulties in achieving something like this? Because the Shawi from its origins has been an autonomous group. But its autonomy through the ages has been undermined by state projects to hispanicise the Shawi, to incorporate us into national economies which are detrimental to the environment, to our culture and spiritual life. It's now clear that conversations about the autonomy of the Shawi people and other indigenous cultures has to be done in a national context. We don't know if the state will accept autonomous indigenous governments.

I: *I think that identity and autonomy are related. What do you think? What activities do the indigenous people of the Amazon do to maintain their identity?*

R: We build our identity from the earth. For example, the place where we were born. To maintain that identity, we have celebrations to honour the earth, because she has produced food to keep us alive. That's

> why she has to celebrate with us. This is how we form our consciousness. We say, "Let's drink manioc beer and dance on our land, so that our land also drinks with us." And we also feed the land with manioc. And we do a ritual in which we thank the earth. And of course, there is reciprocity, because in the end the earth also eats us. As if it were a mother, the earth shelters us, embraces you in the end. The earth is the best mother. That's why we feel the earth with our body, with food, with songs, with dances, with rituals. There are mothers who often reject their children, but the earth never abandons you.

Based on thoughts like these, we keep building our identity. We believe that we should work to reinforce and give continuity to our identity so that we can continue to live. Because if we don't do that, the Shawi culture in fifty years won't exist. There will be no language, and we will have eliminated our thoughts and ourselves. And that would be the most shameful thing: the elimination of ourselves.

Identity is also built in the daily life of the community. Children and young people learn from the elderly. Dads and moms tell stories in the evening and at night. They share the stories and knowledge that our grandparents have left us. We tell many stories when we go to our chacras and mingas [communal work]. When we build our homes, we learn from the nine-banded armadillo [*Dasypus novemcinctus*] who taught us how to weave a thatched roof. We learn from the yellow-rumped cacique birds [*Cacicus cela*] who taught us how to make hammocks. So, in our forests we remember how, back when they were people, these animals were the ones who taught us how to build.

Many stories reflect the way we learn from nature. We also strengthen our identity through our educational institutions. For example, the FORMABIAP Educational Center in Zungarococha where I work, which is on the outskirts of Iquitos, offers an intercultural program that trains bilingual teachers in the Peruvian Amazon. This programme has been inspired to train our own teachers. At FORMABIAP, some of teachers are wise elders of diverse indigenous cultures of the Amazon who teach there and share the knowledge that comes from the forests and rivers.

Because indigenous people have been excluded entirely from the politics of teacher training in the country, we have partnered to defend our rights. One of those rights is education. Good quality education. Here my job is to teach them to write in the Shawi language so that they can see the world from the perspective of Shawi culture. Students learn Shawi grammar with Iris Barraza, who is a teacher and linguist. With wise teachers from other ethnic groups, they learn about the cosmovisions of these other cultures.

Young indigenous students who graduated from public high schools come to our cultural centre in Zungarococha with the mindset that our languages are inferior, that our languages don't produce knowledge. These young people arrive with a big inferiority complex, as if they were influenced by former Peruvian President Alan García who once said that indigenous peoples are second-class citizens. But education helps us to understand what has happened to our indigenous cultures. Because you have to tell the world that we are human beings like others. What has happened to us?

At FORMABIAP, we have a library to help us understand history, to read, analyse and reflect. We also have the oral sources of our grandparents. I believe that in order for our Shawi culture to improve, the knowledge and stories that come from these oral sources should become part of the academic programmes of elementary and high school education. Because in these stories and myths are recorded scientific knowledge about plants, animals and the land which are important for the life of Shawi culture. This way we will build and strengthen our identity. We think that these attitudes help us to develop awareness. It's not about someone saying, "Look at how great indigenous people are," but rather it's about making sure that knowledge and culture contribute to our own communities, to Peru, to the Americas, and to humankind.

I: *At present, the countries that lead so-called "Western culture" face a crisis of civilisation. What practices and knowledge could the Shawi people and other indigenous peoples offer the world?*

R: Since we work on formulating our original knowledge within the framework of interculturality, we can offer the world our perspective of an ecology born from our religious vision of life. We have a cosmovision based on practices that value and respect the places and multiple species that live on Earth. We think that the life of our people, founded on the relationships that our ancestors had with nature, gave us the possibility of practising "el buen vivir" for many centuries. This is what we want to continue practising. With our knowledge and attitude towards nature, we want to contribute everything positive that can reduce the risk of extinction of humankind and of the earth. In that sense, we want to give support to the world. Faced with the individualism that characterises capitalism and the social life of Western culture, we have the practice of helping each other in our communal life.

I: *Now, if reciprocity in the exchange of knowledge is added to that identity, interculturality is reached. How do you define this practice of interculturality?*

R: If we talk about people, interculturality is the individual who handles knowledge or techniques from others, which allow that person to improve their life in terms of agriculture, fishing, raising children. It's applying all of these techniques and knowledge to improve your life. Then, this person could live better by being bilingual, that is, by speaking our language and another language, Spanish. In this day and age, a person with only one culture, in our view, is very weak, very poor. Poor in knowledge and poor in resources. But an intercultural person already has different techniques and different knowledge.

By keeping their own techniques and knowledge as the foundation, they can improve their standard of living and can contribute to other societies. That's interculturality: the management of different types of knowledge and techniques of different people that can improve your life. A universal person, not only in techniques but also in positive attitudes and behaviour, that leads to "el buen vivir" because good living is not only about having material things, but also about having the spiritual and behavioural side, all of which allows you to live in harmony. That's what we long for. But today, interculturality is not enough. Because interculturality refers to sharing knowledge and connecting only among humans. As I said before, you have to think about everyone's life, about the world as something very interconnected, very interrelated. That is, relationships among animal species, places and humans.

I: *You speak of the importance of being multilingual, of speaking several languages in order to be intercultural, and of the benefits of interculturality. What role, then, does Western culture play in the formation of Shawi culture? How does this interculturality help the Shawi people?*

R: The influence of Western culture on the formation of Shawi culture is evident. One of the contributions is linguistic, I mean that a language that is not ours when it arrives makes our communities bilingual. The Shawi individual becomes bilingual.

Something that Western culture brings through the Spanish language is different political ideas. For example, this benefits us because we learn about and adopt political positions of other organizations at national and local levels. There is a political interculturality. We manage our own politics, but we also understand and participate in local political life. Some Shawi get involved in city governments. We practise Shawi cosmovisions and politics, but we also participate in regional and national politics.

Another aspect of interculturality is Western religions. Through the Europeans who came a long time ago, we received the name of God, which we did not know before. Our religiosity was based on the spirits of nature, and now with the presence of God other religious points of view can be practised. This came to us from Western culture. We accept and practise other religions because of some positive things they have.

Of course, as we open up to other forms of Western religion, we have to be cautious because sometimes the Americans and others bring us Adventist, Evangelical and Catholic ideologies. This means that we no longer believe in nature's spiritual beings, we don't practise our traditional medicine because they tell us it's the work of Satan. Many of our dances and music are also considered diabolical by these foreign religions. For example, Juan Carlos, because of the influence of Christianity, Shawi women no longer sing songs about nature; they no longer sing their songs in which they compared plants and animals with people. Now the songs talk about God. We don't want the religious beliefs of others to impose themselves over our religious view of the world and of nature.

On the other hand, some positive aspects of this interculturality and bilingualism are that we can read texts from other countries in Spanish and learn about the lives of other peoples. There are literary works from Peru that help us understand our own reality, past and present. For example, the works of Inca Garcilaso de la Vega, Ciro Alegría, José María Arguedas. We also learn from human rights texts that come from other countries. So, then, in addition to having our oral literature, we can learn about other realities, through other languages and other means.

I: *What is good living and happiness for a Shawi person, and how is that good living built among indigenous Amazonian cultures? What role do work and communal activities play in that experience of good living?*

R: First of all, good living and happiness mean being in harmony in the place where you live, with Mother Earth who gave us life. Happiness for Shawi people today is healthy living, a life full of positive energy, a healthy life. That would be to practise good living.

According to our ancestors, good living meant those times when humans could live in peace, in good conditions, when they had a good relationship with the beings of nature, the forest, the land, and the river. As I say, when relationships were harmonious. There was nothing to worry about, there weren't as many diseases. It was when there was an abundance of animals, fish and wild plants. We used medicinal plants and there was all the wood for people to build their houses, their canoes. There was pure air, pure water, and pure soil. There was no smoke in the atmosphere, you could breathe well. That was good living in the past.

Now there are contaminants in the water from oil and the garbage people throw in the rivers. With the destruction of the rivers, the entire balance deteriorates and breaks, bringing about a shortage of food, of fish, and with it come illnesses and diseases. There's impoverishment and malnutrition, and then there's no good living because you're worried about everything. What am I going to do with my son? What am I going to do with my husband, with my wife who is sick from the contamination? In many cases, you have to go to the posta, which is the health centre in the community. But sometimes there are no medicines and there is no money. People die without medical attention. So there is no happiness anymore, happiness is gone. Life is harder for those who live in the city because indigenous people, if they are hired to work at all, they are paid very little, they have to pay taxes, buy food, pay for water, electricity. There's no happiness.

And of course, I'm not saying good living can't be brought to the cities. It is possible, but only with fair working and living conditions. As for the relationship between work and good living, which you asked me about, we differentiate ourselves from mestizo people. We work so we can have celebrations. They work to earn and accumulate money individually. I believe that we Shawi work to eat and to celebrate. Because to have a party you have to work. We have to produce manioc and sweet potatoes, and sometimes we ask for help with our chacras. But that work isn't paid. It's for our benefit, for the whole community. Just so that we can be well. For us, to celebrate is to share food with others, to dance and sing, to share knowledge.

We also work in mingas, where we collaborate with each other. This is how we help each other. For example, a minga is an activity that is organised around a certain need, like when there is a need to prepare a chacra to grow manioc, plantains or other foods. To do this work, we ask for help from family and friends and we give them manioc beer and food.

There are roles in the minga. The women are in charge of preparing drinks. The men go hunting and fishing to get the meat. And when we are in the minga, the women offer the manioc beer and the men serve the meat. The participants are well cared for. They all fulfil their role. For example, spiders, wasps and snakes have the role of defending themselves and attacking. The spirits also participate. The role of the spirits is to retreat, to leave, so that people can do their work.

In a minga, thirty or more people can participate. This makes us very different from the mestizos. Because mestizos work to earn money, so that they can buy their drinks and their food. But for us, because of our work we have ceramics, hammocks, blankets, canoes. We have all of this because of our own work, because of what we do it. Before, in Shawi culture there were no bosses. In mestizo society, there is someone who employs others, sometimes paying only minimum wage while accumulating wealth. We are different from mestizo people.

I: *You mentioned that there are other external elements of the non-indigenous world that threaten "el buen vivir." How do global capitalism and large-scale development plans in the Amazon ultimately affect the lives of its inhabitants?*

R: A lot of it has to do with the exploitation of the Amazon by companies and people that come from elsewhere. All this causes problems like hunger. In Shawi communities today, there isn't such an abundance of things to eat. Not even here in the city do we eat three times a day. In our communities it's worse. Because resources are scarce in the forests and rivers, nowadays in Shawi communities we say jokingly, "One Shawi eats three times a week."

I remember when I was a child there was an abundance of good-sized fish in the rivers, now we don't catch as many and the fish are very small. Before, in the jungle there were collared peccary [*Pecari tajacu*], deer, white-lipped peccary [*Tayassu pecari*], spotted paca [*Cuniculus paca*]. But now, when a hunter goes out, he no longer finds these animals. What has happened?

What happened is that before, the forest was complete. But now, the forest is no longer whole, that is, it's not complete. Because loggers came and cut down lupuna [*Ceiba pentandra*], mahogany [*Swietenia macrophylla*], tornillo [*Cedrelinga cateniformis*], and leche caspi [*Couma macrocarpa*] trees. What has happened? This destruction has been occurring for some time, because over the years humankind began to develop rapidly. In some countries, they started to need the wealth of other people and lands and what nature had to offer in other parts of the world. And then the Amazon became like a supermarket, like a bank where they could come and take resources in great quantities.

So the countries that still need these resources to accumulate wealth and continue their development are the ones that come to exploit the land. And each time that Western industries have needed some resource or raw material, companies, contractors and other agents have come and taken it from the Amazon. They took a lot of lupuna trees from our land to make plywood. There are almost none of those giant lupunas left because they already logged everything. They also took out a lot of mahogany and cedar. Some members of our community sold timber because they had become contaminated with the ideology that you have to sell wood because there's no money to live on. Because the idea of monetary need had already entered the communities some time before. Little by little, they forced us to participate in the market, in the national monetary system. By then we couldn't avoid it. Many times someone would speak up in the communities and say, "We shouldn't sell the trees." But people didn't listen to them. In this way, by helping to destroy our forests, we are doing away with ourselves.

Just imagine, Juan Carlos, do you know how much they paid for a standing cedar tree when I was a teacher in the community? 20.00 soles. And from that tree they made between 50 and 100 boards. An ordinary board is worth 17.00 soles. The community members from that time didn't understand, because they received gifts: a pot, a gun, a tape recorder, a television set.

Some of these things aren't useful to them. Because the recorder gets damaged, at least the pot is useful for cooking, but most of these things don't produce anything. At least the gun is used for hunting. But the other objects don't produce anything. They don't guarantee that life will continue as before. So, they have given away the trees in exchange for objects that don't multiply themselves, that instead have led to destruction. Some members of indigenous cultures accept that these companies come. They say, "When the companies come, we can sell them plantains, pigs and chickens. We are going to have money, and with it we'll be able to buy tin roofs, outboard motors and pots." It's true, we can earn some money. But, as I told you in our recent conversations and interviews for the documentary *El Río*, there is also the destruction of the forest and rivers and the loss of human life, which is irreparable. Because of these companies, many species have been lost. The government promotes some of the reforestation. But a reforested area is not a natural habitat. The birds, where would they go to reproduce? Many species only reproduce in their natural habitat. The money that the companies are going to earn won't benefit everyone. It will only benefit those big companies. Then there will be a problem. We will find ourselves in a crisis in the future. On the other hand, the oil companies have increased the polluting and poisoning of rivers and forests. Oil is destroying the Amazon. Because many of these extractive activities require the use of chemicals that are bad for this Amazonian soil. If we consume the polluted water, we get sick and die. Also, they want to sell the water. Now they want to build dams on the great rivers in the Amazon that for us are like the blood veins of our bodies. They want to sell the water and build industrial shipping channels.

Many believe that the large amount of fresh water in the Amazon could be taken advantage of on the market. But damming and diverting the rivers would cause drought. There would be problems of water shortage because water is what allows for reproduction. Thanks to the great floods, fish and all kinds of reptiles reproduce. But if we divert the rivers, there will be less water. Imagine the amount and flow of water that the Amazon and other rivers have, the way they flood every year, helping all the fish reproduce and helping to fertilise the fields. As I told you before, the river is like the blood of a human being. Diverting the flow of a river would be like diverting a person's blood to another place where it shouldn't go. Blood circulates for the proper functioning of the heart, kidneys and other organs that are an important part of the functioning of the human body. If we diverted a person's blood, we would almost be killing that person. And that's what they want to do with the rivers. We are forbidden to murder, yet by diverting the rivers, we would be murdering our Mother Earth.

I: *What do you think about the future of the Amazon? What can be done to remedy the problems of the Amazon and how these affect all the people in the world?*

R: I think if we Amazonians don't take care of the forest, others won't take care of it either. The Shawi people agree with the need to protect the environment everywhere on Earth, as expressed by the United Nations, the Pope, professors, researchers, and many others around the world. Because the earth is the nest of humanity.

If we don't take care of it, where are we going to live? Where will our children, our children's children live? In this sense, we agree with the ecological discourses of international organizations. We come into conflict with anyone who wants to extract resources irrationally. We have to educate children and people in general in the knowledge of plants and animal species so that we don't kill Mother Earth.

With regard to the Amazon, many scientists say that at the rate we're going, it will be destroyed, there will be fewer and fewer trees and more droughts will come. I also see this and much more through the visions that the sacred plants give me. The Amazon will be like a desert with plants very far apart from each other, like how foresters plant trees, five meters apart. There would no longer be this thick forest with a variety of animals, vines, palms, fruit trees, and medicinal plants everywhere. This will happen and the entire Amazon will be reforested with varieties of timber trees. Outsiders will come and only plant what matters to them, what produces money. The rest of this biodiversity will have disappeared. This is how the Amazon will be if we don't take care of it.

Seeing all of this, and hearing everything the scientists say about the fact that the Amazon influences the rest of the earth, it seems to me that the Amazon should be of interest not only to the Amazonians, but also to all Peruvians, Americans and all inhabitants of the earth in general. Everyone must make an effort to reduce the effects of the modern world in the Amazon. We must all strive to minimise the impact of climate change in the Amazon. The goal for the creation of nature by Cumpanama, our creator, was to use it for the benefit of all. But it seems that we do not follow his advice and his desire, but rather the need for economic growth in Peru and around the world.

What matters is how to make more money in the world. In that endeavour, we will destroy the Amazon. I believe that human beings are very creative, they have created a lot of technology, but they also use it for destruction. I would beg humanity to build wisdom and knowledge of techniques to maintain this harmonious relationship between people and nature. Our goal is for life to continue to exist, for the planet to continue to exist so that it will allow us all to live well and achieve a full life. I believe this is what we need.

PART 2

Interpreting Eco-visions

CHAPTER 4

Environmental Imagination and Wonder in Beatrix Potter

Lorraine Kerslake

Abstract

Drawing inspiration from Rachel Carson's paper 'Help your Child to Wonder' (1956) and Lawrence Buell's seminal term "environmental imagination" (1995) referring to the reader's ability to experience a sense of connection with the environment and a deeper understanding and appreciation of the natural world, this chapter looks at the role that children's literature can play in imagining a more sustainable society and the importance of instilling a sense of wonder to the natural world in children. In order to question the role and power of children's literature it highlights the contributions of Beatrix Potter to the study, protection and writing of nature and looks at Potter's own literary ecology, including her engagement with nature and the sources of her inspiration. By focusing on the figure of Potter and showing how her own engagement with the natural world formed the children's tales she wrote and illustrated, this paper raises key issues for ongoing debates within ecocriticism and environmental humanities such as Potter's use of anthropomorphism and the role of imagination in bringing about change.

Keywords

Beatrix Potter; animals; ecocriticism; Rachel Carson; sense of wonder; imagination

> If I had influence with the good fairy who is supposed to preside over the christening of all children, I should ask that her gift to each child in the world be a sense of wonder so indestructible that it would last throughout life, as an unfailing antidote against the boredom and disenchantment of later years, the sterile preoccupation with things that are artificial, the alienation from the sources of our strength.
>
> RACHEL CARSON, *The Sense of Wonder* 54

© LORRAINE KERSLAKE, 2022 | DOI:10.1163/9789004501270_006
This is an open access chapter distributed under the terms of the CC BY-NC 4.0 license.

In today's environmental crisis environmental humanities and ecocriticism offer critics,[1] writers and educators a framework with which to recover the importance of the natural world in literature and provide a more holistic view to human's relationship to nature. In this context, the way humans behave is shaped largely by cultural values, including the stories that make up our childhood imagination. As Rachel Carson put forth, the path to building an ecological consciousness lies in responding to the awe of nature expressed by children.

Drawing inspiration from Rachel Carson's paper 'Help your Child to Wonder' (1956),[2] and Lawrence Buell's seminal term 'environmental imagination',[3] referring to the reader's ability to experience a sense of connection with the environment and a deeper understanding and appreciation of the natural world, this chapter looks at the role that children's literature can play in imagining a more sustainable society and the importance of instilling that sense of wonder to the natural world in children. In order to question the role and power of children's literature it draws on Beatrix Potter's life and works and looks at how Potter's literary ecology shaped her attitude to the more-than-human world and asks whether her work might still instill wonder in children today.

Potter's literary ecology (her engagement with nature, the sources of her inspiration, and her political commitment with the conservation and protection of the environment) are at the roots of today's ecocritical theory. Taking Cheryll Glotfelty's definition of ecocriticism as "the study of the relationship between literature and the physical environment" (Glotfelty xviii) where ecocriticism "takes as its subject the interconnections between nature and culture, specifically the cultural artifacts of language and literature" (xix), children's literature would appear to be an ideal domain to explore the intersections of social sphere, nature and imagination.

Given that environmental education begins during the early years of childhood, the stories that we read as children influence the way we respect and value nature later in our lives. In today's society, where children are becoming increasingly detached from nature, children's literature is a powerful means of reconnecting younger generations with the more-than-human world by

1 This work was supported by Proyectos I+D+I Grupos De Investigación Emergentes, Conselleria de Innovación, Universidades, Ciencia y Sociedad Digital, Generalitat Valenciana. Grant Number GV /2020/029.

2 The essay, which was first published in *Woman's Home Companion*, in 1956, was later published posthumously as *The Sense of Wonder* (1965).

3 Buell's term comes from his book *The Environmental Imagination: Thoreau. Nature Writing and the Formation of American Culture* (1995).

creating a sense of wonder. This puts children's writers in a position of singular responsibility, for it is largely up to them to transmit not only cultural values, but to educate and teach respect for our natural world. Children's literature has long proved a hospitable habitat for non-human animals, and the intersections between nature and culture and our relationship with animals have long preoccupied children's authors. Environmental criticism and children's literature often share the concern of environmental imagination (sense of place, affinity to animals, etc.) which has its roots in what we read as children.

In their introduction to *Wild Things* Dobrin and Kidd affirm that "it is critical to recognize that any ecocritical look at children's literature must include ecofeminist perspectives" (10). Indeed, ecofeminism's concern with the domination of nature and mistreatment of animals, parallel to that of people, also takes into account that whilst human imagination reconnects us to animals, we still tend to see them as the Other, highlighting the need to address the questioning of what it means to be human alongside that of other beings in David Abram's 'more-than-human world'.[4] In words of Abram "our obliviousness to non-human nature" (28) stems from our profound anthropocentricism. Like Carson, Abram brings to light the importance of awakening our senses and suggests that there is a synergistic relation between the human and non-human which begins in childhood but usually fades in adulthood through our alienation with nature.

In this context, an ecofeminist reading of Potter's life and work may offer new insights. As Marion Copeland (2004) has commented Potter experienced enough repression Potter experienced enough repression as a woman naturalist to "be drawn to ecofeminism's linking of the oppression of women and the domination of nature" (Copeland 71). Indeed, Potter's own story and her scientific study of mycology show how, throughout history, Western society has systematically shut out female writers from nature sciences. As Carolyn Merchant in *The Death of Nature: Women, Ecology, and the Scientific Revolution* (1980), explains destructive attitudes towards women and nature stem from the emergence of modern science and seventeenth century's scientific revolution which came at the expense of women rights and the environment.

Potter was in fact only one of many pioneering women who crusaded during the Victorian and Edwardian Age to preserve and protect nature and revise evolutionary ideas about the relationships between humans and other animals.

4 Abram coined the phrase 'the more-than-human world' as a way of referring to earthly nature and is used as the subtitle of his book *The Spell of the Sensuous: Perception and Language in a More-than-Human World* (1996). Since then the term has been largely adopted by other scholars and has become a key phrase within ecopsychology and ecocriticism.

Potter's early interest in natural sciences and the reasons why she abandoned her studies in mycology have often been overlooked by critics. For Potter did a lot more than draw the natural world. Not only did she draw with meticulous observation over three hundred and fifty illustrations of fungi but she also grew various fungal spores. She was the first person in Britain to speculate in a scientific paper that lichens are symbiotic life forms, and recorded in detail her observations of algal and fungal properties. In April 1897 and after careful preparation, Potter was ready to present her paper 'Germination of the spores of the *Agaricineae*' to the Linnean Society of London. However, as Kerslake argues (81), the male botanists at Kew Gardens were both sceptical and reluctant to listen to a woman's theory on how fungi spores reproduced. Her paper was offered to the Linnean Society through a botanist at Kew Gardens, George Massee, a member of the Society, but her theories were not taken seriously, and the paper was not recommended for publication.

Undoubtedly if Beatrix Potter had been born a man or a century later, she would have been remembered as a well-known naturalist, but in her time her observational skills in natural history and her artistic talents in mycology were not taken seriously. Luckily for her readers, it was this lack of scientific recognition that made her leave her studies in fungi and turn instead to her animals for inspiration, firstly by illustrating books and making cards and later by writing and illustrating her own little tales. She was, in words of Barbara Gates, "especially well equipped for this enterprise" since "From an early age, she knew her animals as she would come to know her fungi- from a scientific perspective" (231).

Potter was born into a wealthy middle-class Victorian family and lived in Bolton Gardens, Kensington. She was educated at home where she studied books and painting and was looked after by her nanny, together with her younger brother Bertram, until he went to school. She did not socialize with other children and saw her parents at bedtimes and on special occasions. Having little company, she turned to nature from an early age, fascinated by animals, fossils and plants, which she spent hours drawing and immortalising.

Although this aspect of her confined childhood and her apparently tyrannical parents have often thought to be responsible for her solitary nature, given the period and class in which she had been born her restrictive upbringing was similar to that of other children of her age. Given that from the age of five she was allowed to keep her own menagerie including rabbits, lizards, mice, newts, guinea pigs, a hedgehog and snails in her nursery, suggests that it was actually far more progressive than many of her contemporaries.

Both her lonely childhood and self-education influenced her art and work. As a child, much of her inspiration came from her frequent visits to both Kew

Gardens and Kensington Gardens, as well as to the Natural History Museum and the South Kensington Museum (later the Victoria and Albert Museum) where she would quietly observe and draw her first sketches and scientific drawings. In her journal, dated Dec 20, 1895, she describes one of her outings there: "Went to the Museum, very empty and quiet. Studied fossils peaceably, and afterwards the insects again, but investments and a general twitch got too much for me. I never saw anything so fearful as the stuffed animals; I had not been in that gallery for a long time till the other day" (Linder 412). Potter's careful observations and feeling of awe point to her early sense of empathy with nature and the non-human. Her engagement with the Other, informed by her stewardship of animals throughout her life and work, is shown as a bond of respect which she portrayed from an early age. She was both an acute observer and a detached scientist; her major concern was to remain as accurate and faithful to the true nature of animals as possible. Her awareness of this allowed her a privileged understanding of animal behaviour from a different angle from other children's book illustrators and writers of the Golden Age who offered a more sentimentalized view of nature. Furthermore, in the case of Potter, text and illustration are equal components in each of her evocative tales.

As Kerslake argues, part of her artistic talent was inherited from her father who was an amateur photographer as well as close friends to artists such as the Pre-Raphaelite John Millais (82). From an early age she grew up surrounded by art and books; her first encounters with Edward Lear were at the age of four, followed by John Tenniel's *Alice Adventures in Wonderland.* She grew up with Aesop's fables and Scottish folk tales as well as Grimm and Andersen's fairy tales. She read the Old Testament, Shakespeare, Swift as well as her contemporaries Lewis Carroll, Charles Kingsley and Edward Lear and was greatly influenced by the picture books of Walter Crane, Kate Greenaway and Randolph Caldecott. However, as Anne Stevenson Hobbs has noted, "she surpasses her contemporaries in the careful accuracy of her observation, and in her questioning, analytical approach." (23). Potter was, both in her art and writing, basically herself. In her journal entry on October 4th 1884, she comments about her own humble attitude to art: "It is all the same, drawing, painting, modelling, the irresistible desire to copy any beautiful object that strikes the eye. Why cannot one be content to look at it? I cannot rest, I must draw, however poor the result" (in Linder 109).

As a young child Potter's drawing books already showed her artistic talent. All the plants, insects and other creatures Bertram and her brought home soon became models for Potter's fine eye for detail. At nine Potter was already writing about the feeding habits of her caterpillars in her journal. She also studied the breathing system of her newts and frogs and kept records of the hibernation patterns of her hedgehog together with specific details of her rabbits' behaviour. Periodically she would measure each of her animals, carefully keeping record in her notebook. Her own pets were of course models for the inspiration behind many of the characters in her books. Amongst her favourites were her rabbits, especially Benjamin H. Bouncer, who she had before Peter, her mice, and her hedgehogs. Amongst her lizards, dogs, cats, horses, hens, sheep, pigs, guinea pigs, mice, frogs, birds and other beasts, she named over 90 different 'pets'. When they died, she would sometimes boil off the skin and flesh to study and sketch their bones, or stuff them, with the help or guidance of her brother Bertram. She once wrote to her brother, somewhat concerned about his pet bat as it was not eating. His reply gave her specific instructions on how to proceed.

> Dear B,
> I am sorry to hear your news of the bat. If he cannot be kept alive, as I suppose he can't, you had better kill him, and stuff him as well as you can. Be sure to take his measurements most carefully before you stuff him.'
> (Letter from Bertram Potter, Oct 12th, 1884)[5]

The young Beatrix Potter obediently carried out his somewhat gruesome request to painstakingly dissect and stuff the bat and then proceeded to draw him as accurately as possible, with the same scientific detachment and mastery that can be found in her little tales. Graham Greene correctly described Potter as "an acute and unromantic observer, who never sacrifices truth for an effective gesture" (232).

As Kerslake has observed (85) her own relationship with animals show that they are both like and at the same time not like us. Her attitude to nature lacked any sentimentality or sense of squeamishness, and this same matter-of-fact quality can be seen in her own little tales. *The Tale of Peter Rabbit* (1902) begins with a horrifying image of Peter's father's death. When Peter's mother warns Peter and his sisters not to go to Mr. McGregor's garden the reason she candidly gives them is that their "Father had an accident there; he was put into a pie by Mrs. McGregor" (9).[6] This transition of happenings is shown in Potter's first edition with an illustration of a young child at the table and a dog drooling whilst Mrs. McGregor presumably serves the pie in which Peter's father had been baked. Much to Potter's dismay Warne insisted on the image being too explicit and frightening for young readers and removed it from the

5 Quoted her in *Beatrix Potter: A Journal*. (London: Frederick Warne, 2011), p. 6.
6 All references are from *Beatrix Potter: The Complete Tales* (2002).

1902 edition. This opposition between nature and culture together with Potter's own no-nonsense attitude is present at many levels in Potter's tales.

In Potter's art the intertwining of human and animal nature often becomes blurred, creating a sense of oneness. One of the reasons why Potter's anthropomorphic representations are so effective is because of the elements of dark humour and sophisticated irony which often appear embedded in her tales. Young children are often amused by Potter's work because of the fact that rules are broken. Peter, like the disobedient child from moralizing texts of nineteenth-century children's literature, is a familiar figure used to exhort the reader to good behaviour. Except that, in Potter's case, the narrator is on Peter's side. Peter escapes from danger and later drinks chamomile tea from the warmth of his bed. The circular womb-like form of the burrow suggests that he has returned safely to home. Peter's escapism suggests that children should be allowed to explore, play outside and get dirty, contradicting Victorian social restrictions and conventions, anticipating Carson's sense of wonder and the importance of discovering nature. Children know that rabbits don't wear waistcoats but can easily identify with her characters and take pleasure in her sardonic humour which favours Peter at the expense of his well-behaved sisters. Despite the fact she rarely depicts human beings in her tales, her dressed-up creatures reveal as much about human virtues and handicaps as about the natural world.

The popularity of Potter's tales has endured not because the world she depicts is comforting and idyllic as her twee illustrations may at first mislead us to perceive, but because of what lies underneath the surface. Indeed, amongst the more controversial aspects of Potter's anthropomorphism is the need to get past what lies on the surface of her work including the subtle change of size and the metaphor of clothing as well as the smoke-screen of human/animal behaviour, when her animal characters display human actions and act as mirrors or models of human behaviour.

Many of her animals appear both wearing clothes or in their natural state. They act like real animals when they appear naked and become personified and mirror human behaviour when they are dressed. It is Peter's clothing that almost prevents him from escaping from Mr. McGregor's garden. Like Adam and Eve after losing their innocence in the garden of Eden, when Peter loses his shoes in Mr McGregor's garden, he becomes more animal. For Peter, the transition from being clothed to his real animal nature takes place when he loses his clothes and is able to escape.

Although not all of Potter's characters appear dressed, those that are usually wear elegant but uncomfortable clothes that prevent them from walking naturally or do not fit properly. Clothes are usually constricting and symbolise tension, parallel to the attitudes concerning social conventions and the decorum of upper and middle-class Victorians. In *The Tale of Tom Kitten* (1907), as Ruth MacDonald has noted, Tom is dressed in clothes that "make the kitten look like a cross between Little Lord Fauntleroy and Tom Sawyer on his way to church" (100). However, not only do the buttons on Tom's coat burst, and his clothes get dirty, but his sisters also fail to conform to civilized human behaviour by losing their outfits too. When they all return home naked, Tom's mother, Tabitha, locks them upstairs in an attempt to hide them from her guests, explaining that they have measles.

As if to reflect the contradictions inherent in animals wearing clothes, Potter's scenes are simultaneously domestic and disturbing, portraying familiar settings such as having afternoon tea, where one might at the same time find themselves becoming a banquet for others. However, Potter's animals appear at their most disturbing when they act like people.

One of these more grotesque moments that Potter carefully entwines in her tale appears in the sequel to Tom Kitten's story, *The Tale of Samuel Whiskers* (1908), first published as *The Roly-Poly Pudding* (1908). Already at the very beginning of the story, Tabitha Twitchit, Tom Kitten's mother, tells very matter-of-factly, in a sardonic reverse to that of Peter Rabbit's mother, how she caught seven young rats "out of one hole in the back kitchen, and we had them for dinner last Saturday" (179). Potter had recently acquired Hill Top Farm in the Lake District, thanks to the profits of her books and, much like Tabitha, found the old farm overrun with rats when she first moved in. In the tale Tabitha decides to lock Tom, along with his two sisters Moppet and Mittens, in a cupboard to keep them under control. However, Tom escapes up the chimney, only to find himself with Samuel Whiskers, a mean old rat, who gives directions to his wife, Anna Maria, to make him a "kitten dumpling roly-poly pudding" (187) for his dinner. From this scene Potter's illustration contains a gruesome underlying tone. Anna Maria rushes upon poor frightened Tom, with a menacing look, rolling and tying him up with string. Whilst Tom Kitten has lost his jacket and appears without clothes, both Anna Maria and Samuel Whiskers remain fully dressed. Anna Maria then goes to the kitchen to steal the dough, leaving Tom under the floor of the attic.

Potter then presents a black and white image of Tom, facing away from the audience with his head towards the wall, bound by his paws. However, an even more disturbing coloured illustration follows depicting the poor kitten lying on his back, head back and mouth gagged with his paws bound tightly to his body. Potter then elaborately informs us of how the rats "set to work to make him into a dumpling with a rolling pin" (190). Her own sardonic brushstroke comes when Samuel Whiskers then asks with some concern "Will not the strings be very indigestible, Anna Maria?" (190).

Thanks to the noise, that attracts Tom's mother, the rats are discovered, and Tom is rescued and safely returned home. Potter ends the tale by conjuring in the reader's mind more gruesome images, in this case of rats' tails hanging on the barn door, with the following words: "Moppet and Mittens have grown into very good rat-catchers. […] and earn their living very comfortably" (195).

In a similar way to Mrs McGregor's pie in *The Tale of Peter Rabbit*, in other tales, other animals serve as food for both predators and humans. In *The Tale of Jemima Puddleduck* (1908) an elegant Mr Tod deceives the naïve duck to the point that not only does she agree to come to his dinner party but she also promises to bring "Sage and thyme, and mint and two onions, and some parsely" (167) for the stuffing. She manages to escape thanks to Kep, the collie dog and two fox-hound puppies who rescue her. Like the cat and birds that appear in *The Tale of Peter Rabbit* the dogs are not dressed, nor do they have the ability to speak but instead bark and behave like dogs do. So much to the point that the irony of the tale resides in the fact that the puppies end up eating Jemima's eggs.

Together with her underlying irony Potter's satirical eye offers a critical attitude towards the excessive materialism and snobbery of her time, criticising Victorian consumerism. *The Tale of Mr Jeremy Fisher* (1906), mocks artificial emphasis on manners. In the tale, Mr Jeremy Fisher first appears very much as a humanised gentleman frog of leisure portraying the blurring between animals dressed up in human clothing. He is first portrayed sitting outside reading the newspaper with his white socks and small delicate pumps that ironically serve to disguise his webbed feet. Jeremy is a predator himself to smaller animals: his lunch consists of a "butterfly sandwich" (124), whilst for dinner he invites his friends to enjoy "roasted grasshopper with ladybird sauce" (130). However, as Daphne Kutzer notes "if Jeremy is a gentleman he is somewhat out of his depth in the wider world of the pond, a world that contains wild creatures that pose an immediate threat to his obliviously happy, upper-class life" (119). The first of these threats is a water beetle which "came up underneath the lily leaf and tweaked the toe of one of his galoshes" (125), followed by the rustling sound of water rats in the reeves which Potter illustrates, like the beetle and other predators, without clothes and with scientific accuracy. As Kerslake argues (88) the irony of the tale is that it is thanks to his mackintosh, and his denial of his animal nature, that Jeremy is saved: "the trout was so displeased with the taste of the macintosh that in less than half a minute it spat him out again" (127).

However, despite these more controversial aspects Potter's stories are, in general terms, more inclined to naturalism since her creatures are real ones, not just people in disguise. In this sense all her animals have real dwellings: Jeremy Fisher lives by a pond, Peter Rabbit and his family live in a sandbank as rabbits do, although the den has a sink and reminds us of a human home. Other animals that coexist with humans live in human dwellings: Jemima Puddleduck lives on a farm whilst Tabitha Twitchit and her kittens live in an old Victorian house, ridden with mice. Moreover, their animal behaviour is exactly what we would expect: frogs live in ponds and like to get wet, mice make holes and gnaw at things, rabbits sneak into gardens and eat vegetables, cats scratch and play. Their animal homes are *safe* whilst the human outside world is full of dangers. But of course, as in the case of Peter Rabbit, they often prefer to go outside their own natural boundaries and come into conflict with the human world.

In *The Tale of Two Bad Mice* (1904), its main characters end up destroying the upper-middle-class Edwardian doll's house where they find everything for show but not for use. The dolls' lives are an illusion of the bourgeois society they live in and reflect Potter's critique of Victorian false appearances and Edwardian excess. However, read in the light of materialism and the insatiable consumerist society in which we live, much of it still rings true. The tale reflects not only Potter's desire to leave home but her own fears and frustrations about domesticity and independence. As Kutzer correctly observes, in the tale we can see Potter's "absolute delight in bringing disruption, destruction, and energy into a house as stifling and life-denying to the dolls and mice as Bolton Gardens was to her" (71). Written at a time when she was caught in her own difficulties of rebelling against her parents and her wish for independence, the tale reads largely as a reflection of Potter's own domestic upbringing and her family's constraints. At its time of publication, Potter, who was approaching forty, was already making a living from her little books, making it possible for her in theory to leave her parent's home, although of course to have done so would have violated her parent's rules and the Victorian norms of female propriety.

1905 was an important year for Beatrix Potter. She decided to buy a farm called Hill Top in the Lake District, near Sawrey. That same year, during the summer, her publisher, Norman Warne, asked her to marry him. Her parents were appalled, since the son of a publisher was considered an unsuitable match for the daughter of a barrister. Potter had argued with her parents about accepting the proposal of marriage, and finally decided not to announce their engagement. However, only weeks later her hopes were cruelly shattered when Warne fell ill and died suddenly of leukaemia at the end of August. Potter was devastated and took to burying herself in her tales. Consequently, the period that followed was her most prolific.

In 1909 she bought Castle Farm, near Sawrey. In 1913, at the age of 47, she married her solicitor William Heelis and moved into Castle Cottage. That same year she published *The Tale of Pigling Bland* (1913), a tale about a pig who finds love and dances "over the hills and far away" (308), symbolising her own escape to a new life. As Humphrey Carpenter writes it reflected Potter's "exhilaration at her own personal escape, in the year that it was published, from the family fold, and the freedom that her marriage granted her" (296). As Mrs Heelis, Potter was known not as a writer but as a Lake District farmer and sheep breeder. With her profits she bought more estates and farms. In her will she left over 4,000 acres of land, including Hill Top, to the National Trust, so as to preserve the beauty of its countryside. She sought both in life and art to influence her readers to conserve the borderlands and farmlands of the English countryside.

Potter's real aspiration is often overlooked perhaps because it is so obvious and yet so simple as she has put in her own words in her diary in February 1909: "I have always tried to show unspoilt nature in my books. If I have done anything, even a little, to help small children enjoy honest, simple pleasures, I have done a bit of good" (in Lear 309). [7] Like Carson, she sought to influence her young readers to conserve the beauty of their local countryside. Both women shared a similar concern to portray the beauty of nature so that the child's 'sense of wonder' would not fade in adulthood. Potter was happiest not as a writer but later in her life as a farmer in the Lake District when she was actively engaged with nature.

From an ecocritical point of view, as Lawrence Buell (1995) has suggested empathy is key to the development of an environmental ethic. Potter's tales show a clear sense of empathy towards nature, but, as Kerslake observes, despite the fact that she created Arcadian, idyllic landscapes "there is nothing romantic about her tales, for her aim was not to elicit sentimentality in her readers. Her anthropomorphism does not seek an emotional response nor does she foster compassion towards animals, instead she shows respect and encourages coexistence" (90). Potter's tales portray nature as it is for as Margaret Blount reminds us, "Animal life is not happy in the human sense; it is merely neutral [...]. Giving these small animals human qualities is to put them out of reach of inevitable fear, pain and death which is their natural lot" (131). In Potter's little tales' nature is, more often than not, "red in tooth and claw", as

7 Original quote comes from a letter Potter wrote to her friend Kitty Holdsworth, 5 July 1928.

Tennyson puts it, and for the likes of Peter Rabbit, Tom Kitten or Jeremy Fisher, the moral is simply how not to get eaten.

The term of Anotherness, as applied to ecocriticism by Patrick Murphy (2006), is also relevant in Potter: the recognition of the Other as a self-existing entity that is respected, rather than exploited. Otherness is always out of reach. The engagement with the other is shown as a bond of respect, although at times this can cause friction. Take Peter Rabbit and the garden. Peter violates the rules of the garden. Not only does he trespass, by entering the garden, but he steals and eats Mr McGregor's vegetables. The natural balance is altered, but in the end harmony reigns. Potter's way of thinking about human-animal relations and her handling of tension and discomfort reflects themes which can raise awareness of a need for change in the reader on both an emotional and imaginative level.

The roots of environmental imagination lie in the experience that we envision through children's literature in our childhood. The books that we read as children come to be a part of us and shape who we become, influencing the way we respect and value nature later in our lives, for it is in those years of wonder when the child's innate curiosity for the environment is shaped. Potter's unique vision as both a scientist and an artist together with her environmental understanding to preserve nature gave her a special touch that few other writers have conceived. The combination of knowledge, imagination, art and science in Potter's tales fosters respect and care towards the more-than-human world.

By engaging children's literature with our imagination Potter touches on the ongoing debate within environmental humanities about the role of imagination as a key agent in bringing about change and a way to influence our sense of environmental ethics. By instilling in her readers a sense of wonder Potter offers a deeper understanding and appreciation of the natural world.

Despite the fact that Potter's tales are deeply Victorian, they can still awaken the imagination of children in today's society, a society where younger generations are becoming increasingly more divorced from nature.[8] Inspired by Carson and the idea that adults can and should help children discover and enjoy nature, Richard Louv (2005) coined the term 'nature-deficit disorder'[9] to refer to how children (and adults alike) are becoming more alienated from the natural world and to warn of the future consequences this would have.

Questioning where we are today in our understanding about the sense of wonder in young children, both Potter and Carson urge us to use all our senses

8 In 2012, a National Trust report called Natural Childhood revealed the growing gap between children and nature. In 2013, the RSPB published a study which concluded that four out of five children in the UK were not adequately 'connected to nature'.

9 See Louv, *Last Child in the Woods: Saving our Children from Nature Deficit Disorder* (2005).

and explore nature with our feelings and emotions. Anticipating Buell, Potter's tales also put forth the claim that today's environmental crisis "involves a crisis of the imagination the amelioration of which depends on finding better ways of imagining nature and humanity's relation to it" (Buell 2).

In this context, perhaps her tales can help us find ways to strengthen that imagination in today's society and attempt to bring about change. Herein Potter's more-than-human world lies hope.

Works Cited

Abram, David. *The Spell of the Sensuous*. Vintage, 1996.

Blount, Margaret. *Animal Land: The Creatures of Children's Fiction*. Avon-Hearst, 1977.

Buell, Lawrence. *The Environmental Imagination: Thoreau. Nature Writing and the Formation of American Culture*. Harvard UP, 1995.

Carpenter, Humphrey. "Excessively Impertinent Bunnies: The Subversive Element in Beatrix Potter." *Children and Their Books: A Celebration of the Work of Iona and Peter Opie*, edited by Gillian Avery and Julia Briggs, Clarendon, 1989, pp. 271–98.

Carson, Rachel. *The Sense of Wonder*. Harper and Collins, (1965); 1998.

Copeland, Marion. "The Wild and Wild Animal Characters in the Ecofeminist Novels of Beatrix Potter and Gene Stratton-Porter." *Wild Things: Children's Culture and Ecocriticism*, edited by Sidney I. Dobrin and Kenneth B. Kidd, Wayne State UP, 2004, pp. 71–81.

Gates, Barbara. *Kindred Nature: Victorian and Edwardian Women Embrace the Living World*. U of Chicago P, 1998.

Dobrin, Sidney I., and Kenneth B. Kidd, editors. *Wild Things: Children's Culture and Ecocriticism*. Wayne State UP, 2004.

Glotfelty, Cheryll. "Introduction: Literary Studies in an Age of Environmental Crisis." *The Ecocriticism Reader: Landmarks in Literary Ecology*, edited by Cheryll Glotfelty and Harold Fromm, U of Georgia P, 1996, pp. xv–xxxvii.

Greene, Graham. "Beatrix Potter." *Collected Essays*. Bodley Head, 1969, pp. 232–40.

Kerslake, Lorraine. "Of Mice, Rabbits and Other Companion Species in Beatrix Potter's More than Human World" in *Reading Cats and Dogs: Companion Animals in World Literature*, edited by Slovic, Scott, Françoise Besson, Zélia Bora, and Marianne Marroum: Lexington Books, 2021, pp. 79-93.

Hobbs, Anne Stevenson. *Beatrix Potter: Artist and Illustrator*. Frederick Warne, 2005.

Kutzer, M. Daphne. *Beatrix Potter: Writing in Code*. Routledge, 2003.

Lear, Linda. *Beatrix Potter: A Life in Nature*. St. Martin's, 2007.

Linder, Leslie, editor. *The Journal of Beatrix Potter 1881–1897*. Frederick Warne, 1989.

Louv, Richard, *Last Child in the Woods: Saving Our Children from Nature-Deficit Disorder*. Alqonquin, 2005.

MacDonald, Ruth. *Beatrix Potter*. Twayne, 1986.

Merchant, Carolyn. *The Death of Nature: Women, Ecology, and the Scientific Revolution*. Harper and Row, 1980.

Murphy, Patrick. "Grounding Anotherness and Answerability through Allonational Ecoliterature Formations." *Nature in Literary and Cultural Studies: Transatlantic Conversations on Ecocriticism*, edited by Catrin Gersdorf and Sylvia Mayer, Rodopi, 2006, pp. 417–34.

Potter, Beatrix. *Beatrix Potter: The Complete Tales*. Frederick Warne, 2002.

CHAPTER 5

Foregrounding Ecosystems: Thinking with the Work of Helen Mayer Harrison and Newton Harrison

Chris Fremantle and Anne Douglas

Abstract

The approach we take to understanding, whether framed as 'measured, objective and in control' or 'entangled and adapting', is key to the health of the life web and ourselves. The problems associated with the measured, objective and in control version of this, have been identified by artists, philosophers and thinkers including Goethe, Steiner, Klee and Bateson over a long period and came to much wider recognition from the 1970s, with for example the Club of Rome's *Limits to Growth*. Artists create ways of imagining the world that inspire us to feel as well as think. Helen Mayer Harrison and Newton Harrison, known as 'the Harrisons', do so through an ecological approach, as evidenced in their works, including *The Serpentine Lattice* on which this essay draws in particular. Grounded in literary movements such as ethnopoetics and pictorial devices including figure-ground reversal, the Harrisons present us with a fundamentally different way of knowing the world. But where the environmental humanities have tended to reject outright ways of knowing associated with positivism, let alone financialisation of ecosystems, we find quantitative and financial proposals in the works of the Harrisons shaped to provoke us to redirect human institutions to address the health of the life web first and foremost.

Keywords

Art, ecology, epistemologies of the environment, environmental economics.

The Predicament of Mankind, the 1970 text from the Club of Rome, is one of the major attempts to re-think humanity's relationship with the planet. It defines the problem in the following terms: "The source of our power lies in the extraordinary technological capital we have succeeded in accumulating and in propagating, and the all-pervasive analytic or positivistic methodologies

© KONINKLIJKE BRILL NV, LEIDEN, 2022 | DOI:10.1163/9789004501270_007

which by shaping our minds as well as our sensibilities, have enabled us to do what we have done" (5). The text goes on to say:

> However, the experience of the past twenty or thirty years has shown with remarkable clarity that the issues which confront us in the immediate present, as well as their undecipherable consequences over time, may not too easily yield to the methods we have employed with such success in the bending of nature to our will. (5)

This statement frames the ecological and planetary challenges we face as an epistemological issue i.e. in terms of knowledge and ways of understanding.

In what sense is the search for different ways of knowing still relevant? Although seriously questioned in the early 1970s, a positivist epistemology continues to be dominant and has led us to, for example to ecosystems services, attempting to quantify the supporting, regulating, provisioning and cultural services provided by nature to humans. This has evolved into the concept of 'natural capital' that begins to evaluate these services in monetary terms.[1] In the 1960s and 1970s the language of systems was a first step towards addressing the damaging effects of industrialisation on the liveable planet (Burnham 15–16). In current discourses the tendency with some systems approaches to create a false sense of the whole or totality is increasingly challenged in search of alternative imaginaries (Latour 94–98).

What might a different epistemology that is ecologically focused look like? What is the contribution of the arts and humanities, in particular the work of artists, in formulating different ways of knowing? The authors of this 1970 document argue that the analytic or positivistic methods that have underpinned the development of industrialised society won't provide the means in the present and future to address the problems being created.

This challenge was also central to Gregory Bateson's analysis of dysfunctional ways of knowing from the perspective of psychoanalysis. In *Steps to an Ecology of Mind* he proposed that ecological health continued to elude because we had yet to evolve a system in which human civilization and environment came together within a complex whole as knowledge (G. Bateson 347–48), an opportunity for greater interplay between the sciences and arts. For G. Bateson, ways of knowing needed to be based in much richer processes of perception, awe and recognition in our understandings of ecological complexity.

1 Ecosystems services assessment has developed significantly since the publication of the Millennium Ecosystems Assessment. For a history see https://www.millenniumassessment.org/en/History.html and on the development of the concept of natural capital (Missemer).

Such processes would enable us to see the profound interdependence and connectedness of all living things and nature's non-linear pattern of events, as is emphasized by Mary Catherine Bateson in her 1999 'Foreword' to *Steps Towards an Ecology of Mind* (M. C. Bateson xi). In her view perception needed to go beyond the surface appearance of things, to recognize the pattern flow within reality, in particular the non-linear patterns of organic growth. This perception combined with knowledge and experience could lead to new learning, particularly paying attention to feedback loops. Awe, for G. Bateson, was based in being able to perceive the complex interweaving of regular events in qualitative ways, a new form of epistemology that involved aesthetic experience. The collisions between the quantitative linearity of an industrial order and the relational, perceptual quality of ecological thinking were becoming increasingly more evident and problematic (Harries-Jones 159–63).

Picking up on the challenge articulated both by the Club of Rome and G. Bateson, we want to focus on the work of artists and what they might offer, with their processes of imagination and feeling, to an alternative epistemology capable of recognizing the patterns and processes underlying ecology. We are interested in exploring a particular practice—that of the artists Helen Mayer Harrison (1927–2018) and Newton Harrison (b. 1932) (known as 'the Harrisons') to understand how this practice frames and develops ecological ways of knowing, in particular with reference to Bateson's characterisation of an epistemology based in perception, recognition and awe. We have selected the Harrisons' practice because it specifically attends to and evokes underlying ecological complexity and interaction, and is foundational in the field of art and ecology and therefore well tested in regional, national and global contexts (Kagan 283).[2] It is also provocative in terms of knowledge, at times seeming to converge and at others apparently diverge from Bateson's approach.

1 The Harrisons: A Brief Introduction

The Harrisons' work has developed over some 50 years as a partnership between two individuals. It engages two artistic traditions—the visual arts and literature. It is manifest in exhibitions comprising proposals for future ways of living, juxtaposed with current ecological collapse. It is driven by carefully

2 The authors, Douglas and Fremantle, have elsewhere addressed the Harrisons use of inconsistency and contradiction addressed the Harrisons use of inconsistency and contradiction ('Inconsistency and Contradiction') and the poetics of their practice including their use of improvisation ('What Poetry Does Best').

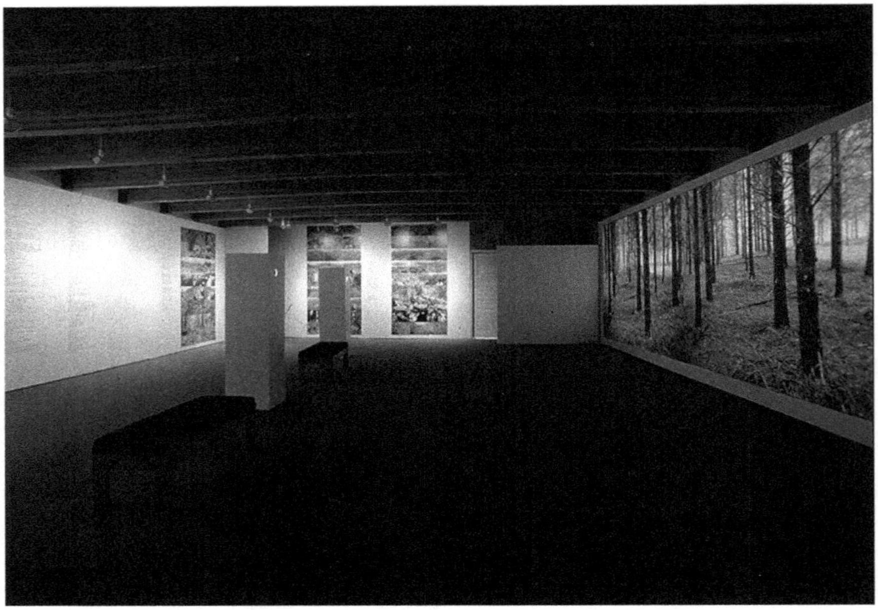

FIGURE 5.1 Helen Mayer and Newton Harrison. *The Serpentine Lattice*, (installation view) 1993. Courtesy of the artist.

framed questions that enable participants to judge what is important in a particular situation. This connects with Schön's idea of metaphor (Schön) as the means to see from a different perspective and in this way to frame issues, and pose questions rather than solve problems. It is context specific in as far as questions arise in particular situations and in specific experiences of environmental change. The Harrisons present this meta-level of understanding to different publics through poetry and image, by telling a story that opens up a new vision. In both content and form they test the boundaries of the visual arts and literature so that their art comes to exist in the world in ways that are different from the artforms (as well as the sciences) on which they draw.

Around the same time as *The Predicament of Mankind* and *Steps to an Ecology of Mind*, the Harrisons made a decision to only create work that takes the perspective of and benefits the environment. They articulate this as "...the progression from an initial decision, made in '69-'70, to do no work that did not in some way look at ecosystemic well-being" (*From There to Here* n.p.). They position the human in relation to the lifeweb, drawing on Fritjof Capra's definition of the 'web of life' in *The Hidden Connection* as a conceptual framework that integrates life's biological, cognitive and social dimensions towards developing a coherent, systemic approach, a science of sustainable life capable of addressing critical issues of our time (Capra vii).

We will argue that the epistemology in the Harrisons' work sharply puts into perspective the conflict between, on the one hand, fragmentation brought about through an economics based in commodification and, on the other, the interests of the commons. For the Harrisons, the commons are air, water, soil, and forests (On The Deep Wealth of this Nation, Scotland). These commons undergird the lifeweb and are most vulnerable in the conflict.

2 Gregory Bateson and Ecological Thinking

Like the Club of Rome's paper, Bateson argues that the twentieth century is characterised by social/cultural pathologies of knowledge that result in irreversible impacts on the health of humans, other living things and the environment. These pathologies are characterized by habits of thinking that look to short-term solutions, maximizing single variables like profit over optimizing the relations between complex variables. Bateson describes how we are trapped in behaviours and values that even in his day were destroying the habitats in which we live and on which our lives depend. Pathologies can emerge as a consequence of simplifying forms of inquiry that trace 'short trains of causality' which can then be easily manipulated. In medicine, for example, cause and effect links in limited circumstances may provide sufficient knowledge for the effective use of drugs or other interventions, but at the same time these disrupt the ecology in unintended ways such as producing parasites immune to antibiotics (G. Bateson 119). Bateson through an investigation of the patterns and processes of living systems proposes a new epistemology, one that locates the mind in relation to its environment (i.e. does not separate the mind from its material base). Bateson's epistemology focuses on the patterns, ideas and processes through which material forms come into being (i.e. not the forms themselves). This new epistemology is also aesthetic: "So by aesthetic I mean responsive to the pattern that connects. The pattern which connects is a meta-pattern. It's a pattern of patterns. It is that meta-pattern which defines that vast generalization that indeed it is patterns which connect" (G. Bateson quoted in Harries-Jones 158).

An ecology of mind is a process in which thinking is an experience *in life* and in the imagination. We respond to information as an experience and (sometimes) self-correct our understandings through feedback loops. This, G. Bateson observes, is a characteristic of living systems from cells to forests, as well as of civilization; a way of knowing that is based in deep systems thinking (109–12).

Mark Engel in his Preface to *Steps to an Ecology of Mind* summarises Bateson's view of the world as follows,

> ...we create the world that we perceive, not because there is no reality outside our heads, but because we select and edit the reality we see to conform to our beliefs about what sort of world we live in. ...Sometimes the dissonance between reality and false beliefs reaches a point when it becomes impossible to avoid the awareness that the world no longer makes sense. Only then is it possible for the mind to consider radically different ideas and perceptions. (6)

Bateson is in fact inviting us to question the self-imposed limitations of our imagination. How might we do so?

3 The Place of the Imagination, Metaphor and Storytelling

Bateson's 'steps to an ecology of mind' offer a number of possibilities including the importance of diversity in maintaining flexibility and resilience, searching for basic continuities that support adaptation and learning from change. He also emphasizes the importance of the story as a form of thought.[3]

---

3 The idea of focusing critically on patterns of thought to nurture ecological understanding is not original to Bateson. Rudolf Steiner developed anthroposophy in Germany in the early 20th century as a philosophy and practice founded in respect for nature. Goethe, as artist and scientist some 150 years before, had already offered the idea that a human being was the finest instrument of observation. By training the body to acutely observe, by paying attention to the surrounding world, we become capable as human beings of forming relations to our environments that are supportive rather than destructive of life (Shotter 151). Goethe formulated his position as a critical counterpoint to the way in which science was becoming harnessed to industrialisation for short-term goals.

> Insofar as he makes use of his healthy senses, the human being is the greatest and most precise scientific instrument that can exist. And precisely this is the greatest disservice of modern science: that it has divorced the experiment from the human being, and wants to know nature only through that which is shown by instruments—indeed wants to limit and demonstrate Nature's capacities in this way. (qtd. in Amrine 37–38)

Paul Klee, the Swiss German artist, musician and anthroposophist drew on Goethe's thinking to form his own approach within the visual arts. In his text *Ways of Studying Nature* Klee says: "For the artist dialogue with nature remains a *conditio sine qua non* (the most essential condition). The artist is human; himself nature; part of nature within natural space" (Klee 15). Klee worked with this idea in drawing and painting, saying in *Creative Confession*: "Art does not reproduce the visible; rather it makes visible" (Klee 7). Painting for Klee was a drawing out of human experience, not a process of representing or illustrating but a way to create life itself through highly tuned forms of observation and imagination.

The Harrisons describe their approach in a way that is strikingly resonant with Bateson's observation about belief and reality above, saying: "Work often begins when we perceive an anomaly in the environment that is the result of opposing beliefs or contradictory metaphors. These moments, in which reality no longer appears seamless and the cost of belief has become outrageous, offer the opportunity to create new spaces, first in the mind and thereafter in everyday life" ("The Harrison Studio" n.p.). The Harrisons offer a way of targeting 'dysfunctional' metaphors that emerge in the way we talk about the places we inhabit. By 'dysfunctional' they mean something close to Bateson's notion of pathologies of cultural construction. The Harrisons flip dysfunctional metaphors by identifying alternative ones that enable a fresh perspective.[4] Their approach is steeped in storytelling: "We hold that every place is telling the story of its own becoming, which is another way of saying that it is continually creating its own history and we join that conversation of place" ('Knotted Ropes, Rings, Lattices and Lace' 14).

The Harrisons not only see 'the pattern that connects,' but harness the aesthetics of pattern to position the lifeweb as foremost and humanity as part of it rather than separate from it. They describe this in terms of a reversal of a figure/ground relationship that draws simultaneously on the conventions of the visual arts (particularly through the work of Joseph Albers) while also drawing on the ways in which scientists might draw pattern from data sets:

> THEN
> A NEW REVERSAL OF GROUND COMES INTO BEING
> WHERE HUMAN ACTIVITY BECOMES A FIGURE
> WITHIN AN ECOLOGICAL FIELD
> AS SIMULTANEOUSLY THE ECOLOGY CEASES TO BE
> AN EVER SHRINKING FIGURE
> WITHIN THE FIELD OF HUMAN ACTIVITY
> (*The Serpentine Lattice* 5–6; format in the original)

This reversal of thinking brings the deep patterns and processes of life to the fore, and repositions human interest as an element of a larger pattern.

4 This connects with Schön's idea of metaphor as the means to see from a different perspective and in this way to set problems rather than solve them. Goto-Collins explores 'the Harrisons' approach in *The Serpentine Lattice* using Schön's concepts (68–69).

4 The Serpentine Lattice (1993)

The Harrisons' work, *The Serpentine Lattice*, from which the quote above is drawn, will function as a key point of reference in this essay. *The Serpentine Lattice* was made at the invitation of Susan Fillin-Yeh, the then Director of the Douglas M. Cooley Memorial Gallery, Reed College in Portland Oregon. It investigated the landscape of the temperate rainforest of the area including the effects of the lumber industry. It was created at a time of increasing awareness of and activism in defence of the forest. It exemplifies clearly how the Harrisons deploy visual and literary metaphor to create a reading of ecological issues. Metaphor becomes a means of positioning these issues within a whole systems approach, reading deeper patterns of ecological damage by taking the perspective of the environment rather than that of human self-interest.

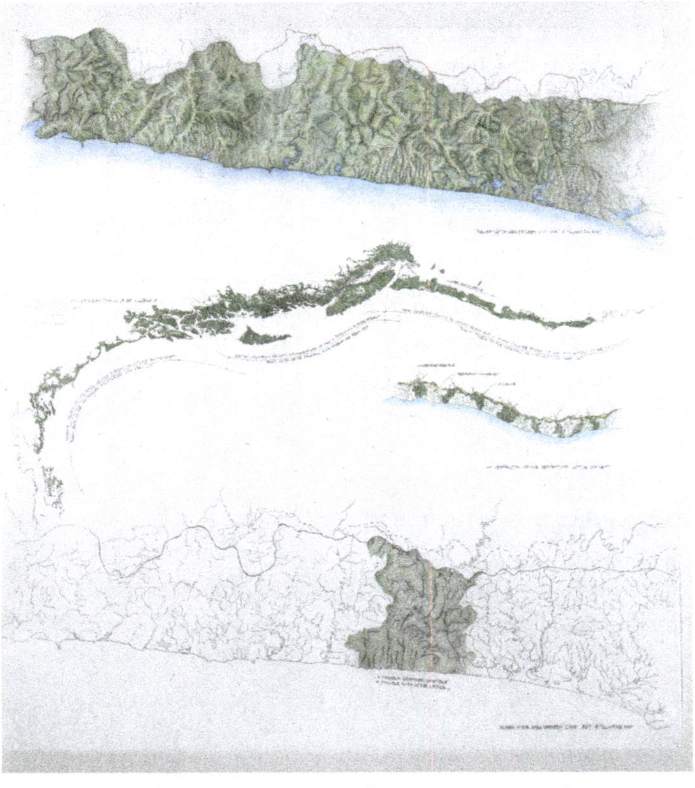

FIGURE 5.2 Helen Mayer Harrison and Newton Harrison. *The Serpentine Lattice*. Courtesy of the artist.

There are several ways metaphor is operating in *The Serpentine Lattice*:

> WE
> BEING GRATEFUL
> FOR THE INVITATION TO JOIN THIS PERILOUS CONVERSATION
> BEGAN TO IMAGINE AN ACT OF RESTITUTION
> YOU SEEING A SERPENTINE
> I SEEING A LATTICE
> WE BEGINNING TO IMAGE NORTH/SOUTH CONTINUITIES
> FROM YAKUTAT BAY
> TO SAN FRANCISCO
> CONTINUITIES THAT WOULD BESPEAK
> THE ECO-POETICS OF THE WHOLE
> ONCE INTERRUPTIONS IN THE FOREST CANOPY
> WERE ANOMALIES
> NOW SUCH INTERRUPTIONS
> AS CLEAR CUTS
> AND TREE FARMS
> AND ROADS
> AND OTHER ARTEFACTS OF CIVILISATION
> HAVE BECOME A NEW NORM
> AND THE RARE INTACT CANOPIES OF OLD GROWTH
> HAVE BECOME THE ANOMALIES
> WITHIN THE MANAGED FOREST
> AN UNFORTUNATE REVERSAL OF GROUND.
> (4)

The Harrisons use the metaphor of joining a conversation as a way to position the art in relation to the on-going life of the place. They use two voices to open up alternative readings of place. This articulates their relationship to the pattern as well as the important elements of the pattern. The landform of this region is serpentine in shape. Logging has consumed huge swathes of what had been the largest temperate rain forest in the world, suggesting a lattice of interruptions. By drawing these together in a composite generative metaphor—the serpentine lattice—it becomes possible to imagine the area as a whole ecosystem and to challenge, if not reverse, the trajectory in which forest was increasingly becoming an anomaly and the industrial extraction the norm. We grasp the destruction (in Bateson's term the 'pathology') at work across different understandings of value: the artefacts of civilisation in the form of clear cuts, roads and tree farms 'disrupt' and effectively displace what once was the continuity in the form of canopies of old growth to the degree that these have become the anomalies within a new system.

The poetry and image together are performative, creating a quality of encounter that is distinctive. This forces us not only to see the internal patterning of an ecology but also to understand how this different way of seeing is itself constructed, through rhythm, skill and in material form. Bateson argues that one of most important characteristics of art is exactly this: art simultaneously offers an experience, a combination of form and content, but also creates the awareness in the viewer/audience that that experience is constructed (G. Bateson 110).

The Harrisons enable us to imagine how the human is nested into the ecological rather than the ecological simply being an 'issue' for the human. This is apparently oppositional to the anthropocentric (self-interested) thinking that underpins current national and international policy. The Harrisons' focus is to pay attention to the ongoingness of life itself, to derive ways of thinking from the complexity of the lifeweb, to confront that ever-present tension within Western thought between intuitive and analytic ways of knowing.

Through this imaginary they construct proposals that are responsive to the complexity of the lifeweb even in its damaged state. These proposals aim to change our way of thinking rather than fix problems. In fact, most often the basic proposal is to stop acting in destructive ways and allow the lifeweb to heal itself. Intervention may be necessary in specific circumstances, but 'getting out of the way' is also important. To draw from *The Serpentine Lattice* again:

> THEN
> WITHIN THIS LATTICE COULD BEGIN
> THE RESTORATION OF THE MORE PRISTINE ENVIRONMENTS
> BY LEAVING THEM ALONE
> BY ENGAGING IN MORE ACTIVE RESTORATION
> ONLY
> WHERE CLEAR CUTTING HAD BEEN MOST SEVERE
> BY CLOSING OFF ENTRY ROADS
> WHEN FINISHED
> A WILFUL GETTING OUT OF THE WAY
> A FELICITOUS WITHDRAWAL
> (5)

The text puts the recognition of the fundamental 'pattern' first: that the lifeweb will restore itself if left alone. Intervention should be shaped and moderated by that understanding.

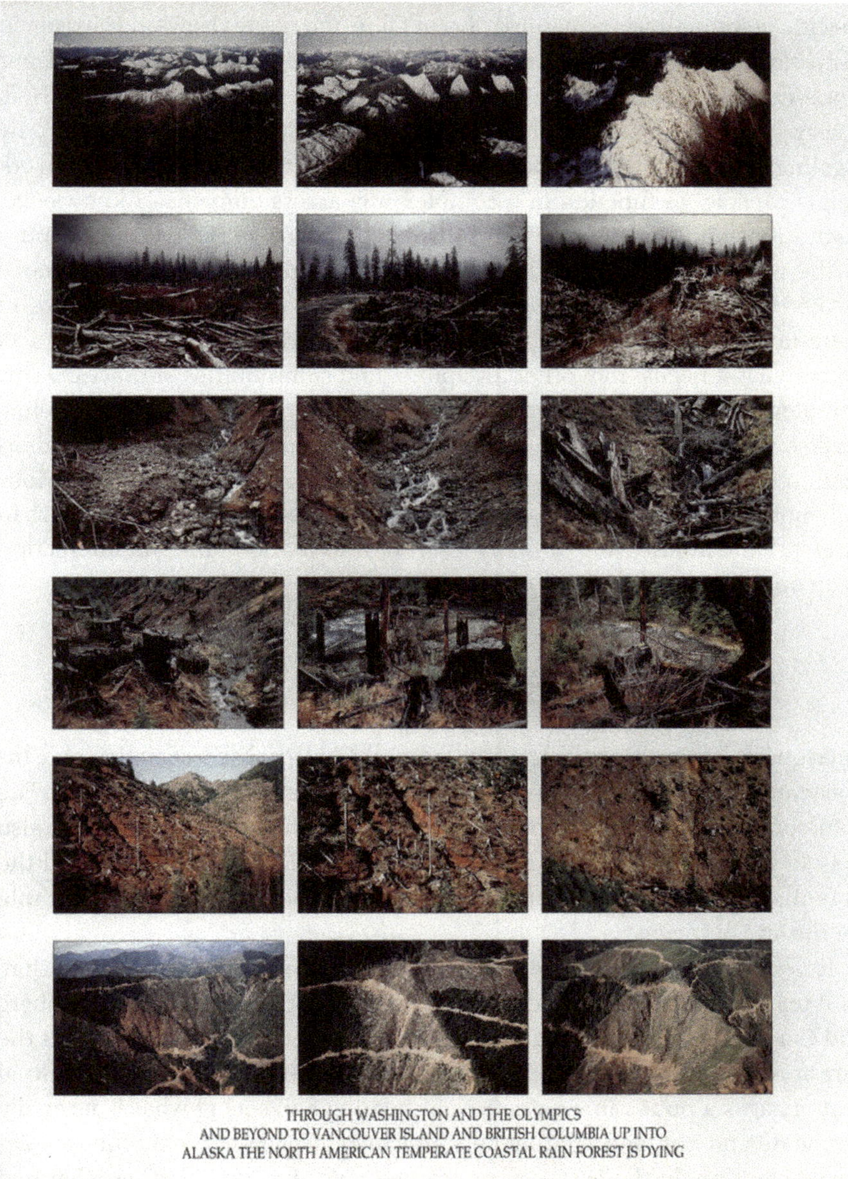

FIGURE 5.3 Helen Mayer and Newton Harrison. *The Serpentine Lattice*, (detail) 1993. Courtesy of the artist. (Original photographs: Trygve Steen).

The use of propositional form in the work is significant in opening up an invitation to imagine the world 'as if', rather than 'as is', indeterminate rather than determined. The propositional is also an invitation to test or verify the correctness, to see knowledge of any kind as relative to ways of knowing at a specific point in time and context. Susan Fillin-Yeh quotes Newton Harrison in her essay for *The Serpentine Lattice* saying, "In the context of the art world, our works do, in fact, behave like works of art. When they're exhibited at City Hall, however, they read as workable proposals in poetic form" (Fillin-Yeh 17). By toggling between scales and between contexts of art and usefulness, this work builds capacity to function in a complex way across domains of knowledge, institutions and organisations and diverse practices without losing identity as the work of art. Being and knowing are entangled together. The practice of the Harrisons does not follow the canonical in either the visual arts or in literature, but synthetically draws what it needs from both artforms. It also draws what it needs from other disciplines and communities of practice. This complexity characterises their approach. They combine reason and feeling while exercising the right to question what might appear self-evident in information exchange by pointing to what Bateson describes as, "…the mysterious polymorphic relations between *context* and *content*" (G. Bateson 124 italics in original), the interleaving of many levels of information flow and interaction in an ongoing state of emergence.

5 The Role of Ethnopoetics

Bateson, as an anthropologist, studied non-industrialised societies and his epistemological arguments are informed by their eco-cultural modalities. The Harrisons, through their association with the ethnopoetics movement, also look to the role of poetry in non- and pre-industrial societies to expand the potential for different ways of knowing through the arts, and a different role for the arts in society.

Helen Mayer Harrison had had a lifelong interest in language, narration, storytelling, and the oral tradition. For her and others (Jerome Rothenberg and Gary Snyder, both poets and anthropologists) ethnopoetics highlights the transformative power of vision, of spoken and performed storytelling, of ritual that engages a public in experiences of hearing and seeing what is normally not heard and not seen.[5] Rothenberg and Snyder suggested that ethnopoetics in the early 70s in the USA was concerned with drawing together what had become fragmented in a heavily industrialised society. By exploring the poetry of non-industrialised cultures and their deeper levels of direct contact with the natural world, ethnopoetics set out to enrich understanding of what poetry

5 In San Diego the Harrisons were close friends with David and Eleanor Antin and with Jerome and Diane Rothenberg, all important members of the movement.

might be in occidental industrialised nations (Snyder 2). Rothenberg writes, "The suspicion came to be that certain forms of poetry, like certain forms of artmaking, permeated traditional societies ∓ that these largely religious forms not only resembled but had long since achieved what the new experimental poets & artists were then first setting out to do" (1).[6] Rothenberg stressed the transformative power of ritual, the importance of rhythm, of movement and of transformation to draw out the deep patterns which both Bateson and the Harrisons are emphasising.

Both Snyder and Rothenberg resonate with Bateson's concern to reintroduce aesthetics as core to ecological sensibility and knowledge in terms of rhythm if not also ritual. Pre-industrial forms of art making, whether the poetry of Northwest American First Nation people (the Harrisons) or Balinese painting (Bateson), appeared respectively as capable of integrating different levels of communication.

This conjunction of poetry and ritual with its role in a different epistemology is also present in the Scottish context (where the authors live and work). In Scotland this goes by the name of Geopoetics and draws on the Celtic Bardic tradition that has, as writer and activist Alastair McIntosh argues, played a vital role in the renewal of rural communities in the Scottish Highlands. They are also the spiritual conduits of the people, opening up fresh channels within the mind and collective forms of consciousness, forming a kind of inner order from which the energies of the material world emerge. For McIntosh poetry is not just the written word, it is a way for finding a different relationship with land. He says of this kind of poetry, "And by 'true' art forms, I mean those that come from somewhere deeper than the ego—those that emerge from the collective levels of consciousness" (n.p.). This is a form of storytelling that is very close to indigenous cultures where the re-telling of process is a means to communicate both knowledge and value.

The Harrisons' work establishes a set of principles framed by two important questions:

- What does knowledge mean to here?
- What opportunity does this knowledge shape?

We have already noted Bateson's emphasis on context, but the second question significantly expands the question of context by opening this up to possible future action. The Harrisons express the potential for action in the form of a proposition. In addition and in contrast to Bateson's focus on qualitative approaches, they draw on the tools of quantification. So if the Harrisons' work

6 ∓ is a symbol used by Rothenberg.

sits broadly within a tradition of ecological thinking (Goethe, Steiner, Klee, Capra and Bateson), and more specifically draws on ethnopoetics (Rothenberg and Snyder), why do we find references to Gross National Product (GNP), descriptions of 'externalities', and proposals for raising taxes on water appearing in their works from the early 1990s onwards?[7] This seems to be a gesture towards not displacing one epistemology with another, but acknowledging the potential for constructive ways of integrating the quantitative with the qualitative within a propositional form.[8] We find GNP appearing in *The Serpentine Lattice* (1993) and it reappears in the *Peninsula Europe* works, I (2000–03), II (2007), III (2008) as well as in *Greenhouse Britain* (2007–09). In the passage below, *The Serpentine Lattice* asks us to see the health of the lifeweb as underlying the health of the economy, in fact as the underlying organisational requirement for the economy.

> FOR INSTANCE
> IF THE GROSS NATIONAL PRODUCT
> IS 5.7 TRILLION DOLLARS
> AND
> PRODUCING THE GROSS NATIONAL PRODUCT
> IS THE OUTCOME OF EXPLOITING
> THE GROSS NATIONAL ECOSYSTEM
> AND
> THE GROSS NATIONAL ECOSYSTEM
> IS NOT INFINITELY RENEWABLE
> THEN
> IT IS NOT DIFFICULT
> TO IMAGINE THE GROSS NATIONAL PRODUCT
> SHRINKING

7 The Harrisons use everyday terms such as Gross National Product which, along with Gross Domestic Product, have more recently been criticised, for example by Arundhati Roy e.g. in *Capitalism: A Ghost Story*, (2014). Suggestions of replacements such as Gross National Well-being or Happiness have gained some traction, and ecosystem health indicators are increasingly discussed (O'Brien et al.).

8 To try and understand this it is worth bearing in mind Gary Snyder's observation that an anthropologist should only engage in explaining another culture if they can explain their own culture's economics. Quoting Dr Jack Stauder, Snyder says, "…if you're going to be an anthropology teacher you should also be able to teach your students the dynamics of their own culture, at least in the critical area of understanding imperialism and capitalism. If you can't communicate that to your students, then you've got no business talking to them about the Xingu. If you can't explain the banking system, well, where are you?" (4).

IN CONCERT WITH AN OVEREXPLOITED
LESS PRODUCTIVE
GROSS NATIONAL ECOSYSTEM
HOWEVER
IF
AS A FORM OF RECYCLING
WE TAKE 1%
OF OUR GROSS NATIONAL PRODUCT
AND ESTABLISH
AN ECO-SECURITY SYSTEM
NOT UNLIKE
OUR SOCIAL SECURITY SYSTEM
THEN
ROUGHLY 57 BILLION DOLLARS
BECOME AVAILABLE YEARLY
FOR RESTORATION/RECLAMATION
(7)

The artwork proposes a new tax on the Gross National Product to pay for restoration and reclamation. How is this not just financialising the lifeweb in the positivist ways that Bateson is critical of?

The Harrisons integrate the financial way of seeing the world into their overall perspective from the health of the lifeweb. Specifically, they draw out the problem of longterm costs and energy debt in terms of timescale, not just impact:[9]

FOR INSTANCE
WHO WILL PAY THE LONG TERM COSTS
OF CHANGING WEATHER PATTERNS
AS THE MOISTURE RETAINING PROPERTIES
OF THE LARGER TREES
DISAPPEAR
AND THE CARBON SEQUESTERED BY THE GIANTS
IS RELEASED INTO THE ATMOSPHERE
AND THE OXYGEN REPLENISHING PROPERTIES
OF THE QUADRILLIONS OF LIVING NEEDLES

9 This issue is now framed using the terminology of 'externalities', costs on a third party beyond a transaction. The clear-felling of the forest reduces the health of the river stream habitats reducing their water purifying functions as one of several examples in the text.

ARE SUBTRACTED
AND THE GRANDEUR
OF THE ANCIENT
FOREST SYSTEM OF THE GIANTS
DISAPPEARS
...
AFTER ALL
THIS LONG TERM ENERGY DEBT
COMES DUE
IN THE NEXT GENERATION
WITH THE TURNING OF TENS OF THOUSANDS OF SQUARE MILES
OF BIOLOGICALLY PRODUCTIVE LANDS
INTO FUNCTIONAL DESERTS
AND THE ELIMINATION
OF PRODUCTIVE ECOSYSTEMS
FROM OVER ONE HUNDRED THOUSAND MILES
OF RIVER-STREAM HABITAT
AND THE WATER PURIFYING PROPERTIES
OF THE WETLANDS
DISAPPEAR
(6–7)

What is particularly important about their framing of these externalities is the duration in the shape of a long-term energy debt. This energy debt is coming due (we are after all functionally a generation on from the making of this work). In addition, the scale of the debt needs to be understood in terms of the hundreds and thousands of miles of river stream habitat not providing fresh water, or the quadrillions of needles not replenishing oxygen (i.e. not contributing to the lifeweb on which we and all living things depend).

In the intervening years the issue of externalities, and the wider development of environmental economics, has become better known and clearer. As noted earlier methods such as Ecosystems Services Assessment and Natural Capital Accounting have become increasingly mainstream. The Harrisons make proposals for taxation as a means of addressing these externalities, intending to reinvest in the health of the lifeweb. In particular, in a later project, *Peninsula Europe* (2001–03), we find the following:

> You said, “think of the waters. About 1430 billion cubic meters of waters fall on the icon. At least a thousand billion flow downhill. The rest remains

> in the mountains because of percolation and evapo-transpiration. So we are looking at about a thousand billion cubic meters of water yearly."
>
> Then I said, "Everybody pays at least one DM [Deutsch Mark] and sometimes two or three times that downstream for clean water. So we are looking at a trans-peninsula expense pattern of perhaps 2 trillion DM yearly assuming all downhill flow is used at least once. As the gross transnational product appears to be about 16 trillion DM, a rough calculation suggests that a modest water tax of about 3/5th of a per cent would yield close to one hundred million DM a year. Projecting this over a 15 year period, that amount of money would go a long way to putting this system in place. Once in place the system would move to self-maintenance and therefore the costs would drop significantly." Then you said, "It looks like a very modest water tax for the good of the whole." (*Peninsula Europe* n.p.)

This sequence—the positioning of GNP within the lifeweb, the framing of the scale of externalities and the urgency of taking these into account as well as the proposals for taxation capable of addressing the health of the lifeweb—all form a coherent way to understand our own economics as part of ecological thinking. The Harrisons bring to the fore all the critical aspects of environmental economics. In place of extraction, they propose the introduction of financial feedback loops which did not exist, creating mechanisms to give back to the web of life. They do so in a way that is both comprehensible and poetic, while also unapologetically drawing our attention to crucial patterns of relationships and dynamics.

What does the Harrisons' articulation of large-scale numeric/financial propositions achieve?

The Harrisons address this through art and its power to provoke the creative imagination of both artist and viewer/reader. They say, "Big figures were an empowering aid to thought, invention, improvisation, and play" (*The Time of the Force Majeure* 316).

In other places they talk about defining 'a field of play' as being the same as a painter defining the scope of the composition ('Public Culture and Sustainable Practices' n.p.). Their maps are compositions just on a different scale from most paintings. Their use of figures achieves the same function of composing at scale, creating an image of ecosystems that complements and adds a dimension to the maps in the experience of the viewer. The numbers are another means of creating images in the mind, and even of provoking new thinking through proposing new patterns of organisation between the ecological and the economic.

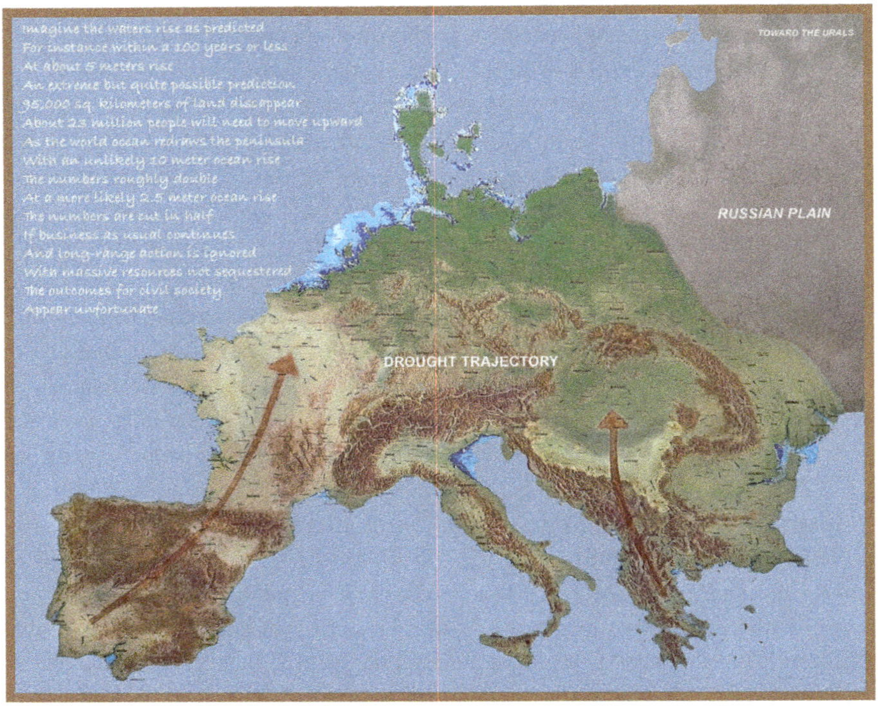

FIGURE 5.4 Helen Mayer Harrison and Newton Harrison. *Peninsula Europe II: Drought Trajectory.* Courtesy of the artist.

Moreover, just as they push back the human elements (roads) and bring forward the natural elements (rivers and mountains) in their creation of map images, the figures they select to focus our attention on construct scale and proportional relationships in the manner of design and architecture. These consist largely of areas of land, volumes of water and quantities of organic waste. In this way the Harrisons create a vivid image of value between these natural elements and human concepts such as Gross National Product. In addition, the numbers they use ask us to pay attention to the integrity of the ecosystem in contrast to commodification practices that use numbers to increase fragmentation for monetary profit. In the case of both *The Serpentine Lattice* and *Peninsula Europe*, their composition is at a transnational scale (i.e. the scale of ecological pattern rather than of human political organisation).

It is worth bearing in mind that the Harrisons start projects with the questions "How big is here?" and "How long is now?". These questions ensure that the "field of play" is in both dimensions of time and space and neither limited nor determined by existing political or economic assumptions ('Public Culture and Sustainable Practices' n.p.).

If we consider the current frameworks used for economic valuation, this is often framed in three stages: "First, the way a decision will influence the environment needs to be understood (qualitative assessment). Second, the change in the environment and the related benefits need to be measured (quantitative assessment). Only then can the third step of valuation in monetary terms take place" (Ozdemiroglu and Hails 5). Interestingly, the Harrisons evoke all three of these elements: qualitative, quantitative and monetary. But their starting point with questions of scale and time ensure that the focus is not on a decision, but rather first on the lifeweb's integrity (so far as we can understand that). The significance of this point is further reinforced when we consider in answer to the question 'Whose values count?', the Valuing Nature Programme, for example, reduces scale to the national political (Ozdemiroglu and Hails 8). The Harrisons always attend to the lifeweb scale (the temperate rainforest stretching from Yakutat Bay to San Francisco), rather than the political boundary (four different states—California, Oregon, Washington, and Alaska—in the US, and British Columbia and even part of the Yukon in Canada).[10]

Finally, the Harrisons' proposals, as we have shown in the examples of *The Serpentine Lattice* and *Peninsula Europe*, are focused on introducing new feedback loops in the first instance to the benefit of the lifeweb. The proposal for a tax on water made in *Peninsula Europe* (above) is to fund the restoration of the high grounds, increased forestry and biodiversity, which will in turn produce more clean water (and probably less flooding). Their proposal for taxation becomes a mechanism which serves the lifeweb as well as human needs.

6 Moving to a Conclusion

This essay opened with the Club of Rome and Bateson's critique of forms of epistemology that are increasingly damaging the global environment. The Club of Rome was a meeting between disciplines and international/global responsibility for governance in relation to the environment. Bateson's critique emerges out of the discipline of anthropology and is not new, but is threaded through Western thinking, including philosophy, art and the natural sciences since the emergence of industrialisation. Bateson focuses on practices of quantification as a manifestation of a deep problem in the way the West imagines

10 For further discussion of the difference between national scale interest and lifeweb interest in the context of *The Serpentine Lattice* see Goto-Collins (67).

and acts upon human relations with the non-human world. In this he follows Henri Bergson.

In his *Introduction to Metaphysics* Bergson proposes two ways of knowing the world. The one moves around an object constructing a sense of the whole through partial views (e.g. through weighing and measuring), a form of analysis that through fragmentation attempts to understand the whole. The other, which uses intuition too, seeks to grasp an object from within itself and in its living mobile state. Any attempt to construct our sense of the whole by setting one concept against another, fragments our understanding and is an illusion, he argues. Bergson invites us instead to focus on movement and in this way to intuit the inner life of objects in the world through duration (i.e. viewing life in the making, in a continuing process of formation and adaption).[11] Measuring and weighing therefore conceptually entails stopping the world in its tracks, breaking continuity and movement, interrupting the mobility that is our natural state.

In the essay we propose that the arts, particularly as exemplified in the work of the Harrisons, can offer a different way of knowing that is distinctive and capable of addressing Bateson's call for a different epistemology shaped by attention to the underlying complexity of living systems. We explore this through the practice of the Harrisons whose work has pioneered a new field of inquiry linking the arts with ecology. We note that a practice is different from a discipline. It is entangled with the world, already part of the way life comes to be formed. It is not extracted from or applied in the way that knowledge is imagined in many discipline-based domains. This quality of relationship is resonant of Bateson's understanding of an ecologically focused epistemology. The connectivity of the mind is not limited by the skin, not confined to the interior of a living being but rather an embodied presence in a specific environment that transforms its surroundings through interaction (G. Bateson 326).

The Harrisons' particular approach and way of making art is important because it has evolved over fifty years. It demonstrates the potential for art to propose, evoke and make visible patterns, drawing attention to those that are

11 Bergson's extrapolation of these two ways of knowing is compelling:

> Let us, then, rather, imagine an infinitely small elastic body, contracted, if it were possible, to a mathematical point. Let this be drawn out gradually that from a point, comes a constantly lengthening line. Let us fix our attention not on the line as a line, but on the action by which it is traced...Finally, let us free ourselves from the space which underlies the movement in order to consider only the movement itself, the act of tension, or extension; in short, pure mobility. We shall have this time a more faithful image of our self in duration. (26)

dysfunctional, and offering others that respect the lifeweb. Their art has been tested in a significant number of cultural contexts and in relation to a rich diversity of environmental problems in specific locations. Importantly, it has taken a place at the table of other conflicting public discourses and entered into dialogue with disciplines concerned with the environment, frequently acting as the pivotal point that draws such diversity of thinking together.

In conclusion we would like to draw attention to four key qualities of their approach that we have exemplified in the text and that arguably together offer the kind of radical epistemological shift that Bateson was seeking.

The Harrisons, like Bateson, position human beings as part of the lifeweb and this positioning underpins all their other proposals and revelations. Their most recent work, a cluster of projects in different parts of the world developed through *The Center for the Study of the Force Majeure* (2016 ongoing), goes further in suggesting that human beings can no longer assume a dominant, or controlling, position in relation to 'nature'. We can now only adapt because the consequences of human-induced environmental damage are so severe that the lifeweb is pushing back, becoming this 'force majeure', in the form of three interrelated elements: heatwave, sea level rise and biodiversity loss (6th Great Extinction).

The Harrisons draw us into this different way of knowing through their poetics, communicating complex ideas in ways that we not only recognise, but that also move us. They engage us in experiences that are affective—experiences of perception, recognition and awe. These are the characteristics through which Bateson defines a way of knowing that includes the aesthetic. Knowledge in this form is not information but formative of a way of being in the world and, in the case of the Harrisons, can lead to action.

There have been many arguments for sensory, entangled, holistic ways of knowing as a counterpoint to positivism. The scientist Donna Haraway, for example, draws on the arts and in particular literature, to explore thresholds, known as 'contact zones', between the human and the more-than-human (Haraway 33). The arts are often positioned in opposition to the analytical and positivistic, but artists, including some at the forefront of thinking ecologically such as the Harrisons, have also used manifestations of the analytic and positivistic ways of knowing, what we might simplistically characterise as counting and measuring, within their works.[12] The Harrisons' work as artists, in contrast

12 Other examples might include Duchamp's *3 Standard Stoppages* (1913–14) or Hans Haacke's 1971 piece *Shapolsky et al. Manhattan Real Estate Holdings, a Real-Time Social System*. Duchamp (1887–1968) configures the standard unit of a metre in three different ways, none of which are the conventional straight length, reminding us that measurement

perhaps to Bateson's position, does not set aside such existing forms of knowledge. Instead they confront the implications of current forms of quantification in the way culture and society has used them, exposing the dangers of increased fragmentation and turning existing knowledge towards the development of common goals. They use quantification as a way to evoke patterns through techniques such as proportion and scale, to discern crucial differences and they propose ways that quantification can enable new feedback mechanisms. Specifically, they turn all forms of knowledge towards the overarching goal of giving back to the lifeweb more than we take.

Finally, 'knowing' in this way is mobile. It does not involve conceptually stopping time. For Bateson one of the significant dangers of industrialisation has been the loss of a sense of time as oscillation, the recursive time of organic life. Human beings have, he suggested, internalised the linear, quantifiable time of industrialised order in ways that are deeply embedded in daily life to the point that we have difficulty in extracting ourselves. We have come to believe that human beings can have control over events (Harries-Jones 165–66). In the Harrisons' works they frequently speak of a moment in terms ranging from 25 years to 1200 years (*The Serpentine Lattice* 7 and 3) i.e. extending well beyond the life or control of an individual. The Harrisons talk about "joining the conversation of a place" and comment how "every place is a story of its own becoming" ('Knotted Ropes, Rings, Lattices and Lace' 14). Time in this construction is imagined as continuous and subject to rhythms. They talk about improvisation as a way of imagining cultural systems within ecological systems, as reflexive, unstable and fragile (*The Lagoon Cycle* 37 and 60) and they say, "...any central images that appeared seemed to exist only for a moment and thereafter fade back into a pattern of moments grouped within moments" (*From There to Here* n.p.).

David Harvey, another anthropologist writing in the present, argues that the fourth of the 'seventeen contradictions of capitalism' is the privatisation of common goods, principally land, labour and money, three aspects of life that cannot be 'owned' by individuals but should be shared in common. Land is nature and not produced by people—it can be stewarded, but not owned. Labour, by which he means work, goes with life itself and is not for sale, even

is a human construct that may be imposed as a way of standardizing the world imagined as controllable. Alternatively, the same measure may be experienced as unique and random when interacting with other complex forces such as relative weight and gravity (Wilkins 373). Haacke's piece draws on data, the kind that might be associated with the social sciences and social policy to expose the inconsistencies between market requirements and the social needs of a disenfranchised community.

though it can be remunerated. Money is produced by a banking system governed by the state in the interest of the whole population, not of a few. The main driver of privatisation and fragmentation in capitalism is commodification, enabled by positivism's counting and weighing, though not necessarily entailed by it. Harvey suggests that the current intensification of forms of privatisation are not only fundamentally fictitious but also highly damaging to society, robbing the protection previously afforded by public institutions. Privatisation is a form of appropriation and appropriation implies an act of taking something for one's own use without permission. He shows the degree to which such practices have become normalised and how this is supported by poor public policy, saying: "Once land, labour and money had become objectified, pulverised and broken away from their embeddedness in broader flows of cultural life and living matter, then they could be resutured together under the constitutional rights and laws founded on principles of individual rights to private property" (Harvey 58).

This process is not only fictitious, but also needs to be recognised as illegal. Common resources that should not, and previously could not, be commodified and monetised such as gene sequences, seed banks, oceans through fishing rights and offshore renewable energy developments, weather futures, have now legally become 'privatised'. It is the collusion of governments in developing policies enabling privatisation as the norm for short term gain that in Harvey's view compromises our means to survive as humans.

The Harrisons' work, alongside Bateson among others, exposes with startling clarity the depth to which we have sunk as human beings—the dangerous fictions that increasingly underpin the way we harness resources in support of humans with little care for the environment upon which we depend and the short term 'wealth' being enjoyed in a grossly inequitable way. It has to be said that many of the works we have cited were made in the 1990s and 2000s when structures such as the European Union made it possible to imagine things like transnational watershed authorities which could introduce new feedback loops in the form of financial instruments. Today fragmentation seems much more dominant, ironically brought about by the financial crisis which in turn was caused by a banking system that had ceased to understand its role as a public good.

The Harrisons' current work addressing the 'force majeure' generates the hope that this trajectory may be changed by telling ourselves a different story, a story which reminds us of our deep and enduring connection to and dependence on the lifeweb as a whole; a story which urgently requires us to respect and adapt to the larger forces over which we have no control in a rapidly changing global context.

Works Cited

Amrine, Frederick. "The Metamorphosis of the Scientist." *Goethe's Way of Science*, edited by David Seamon and Arthur Zajonc, State U of New York P, 1998, pp. 15–30.

Bateson, Gregory. *Steps to an Ecology of Mind: Collected Essays in Anthropology, Psychiatry, Evolution, and Epistemology*. 1972. U of Chicago P, 2000.

Bateson, Mary C. Foreword by Mary C. Bateson, 1999. *Steps to an Ecology of Mind*, by Gregory Bateson, U of Chicago P, 2000, pp. vii–xv.

Bergson, Henri. *An Introduction to Metaphysics*. 1912. Hackett, 1999.

Burnham, Jack. *Great Western Salt Works: Essays on the Meaning of Post-Formalist Art*. George Braziller, 1974.

Capra, Fritjof. *The Hidden Connection*. Harper Collins, 2002.

Douglas, Anne, and Chris Fremantle. "Inconsistency and Contradiction: Lessons in Improvisation in the Work of Helen Mayer Harrison and Newton Harrison." *Elemental: An Arts and Ecology Reader*, edited by James Brady, Gaia Projects, 2016.

Douglas, Anne, and Chris Fremantle. "What Poetry Does Best: The Harrisons' Poetics of Being and Acting in the World." *The Time of the Force Majeure: After 45 Years Counterforce Is on the Horizon*, edited by Helen Mayer Harrison and Newton Harrison, Prestel, 2016.

Engel, Mark. Foreword. *Steps to an Ecology of Mind*, by Gregory Bateson. 1972. Jason Aronson, 1987, pp. 6–7.

Fillin-Yeh, Susan. "The Serpentine Lattice: Where You Said a Lattice and I Said a Serpentine and You Said Network the Watersheds and I Said a Game of Go." *The Serpentine Lattice*, Douglas M Cooley Memorial Gallery, Reed College, 1993, pp. 16–23.

Goto-Collins, Reiko. *Ecology and Environmental Art in Public Place: Talking Tree: Won't You Take a Minute and Listen to the Plight of Nature?* Robert Gordon University, 2012.

Haraway, Donna J. *When Species Meet*. U of Minnesota P, 2008.

Harries-Jones, Peter. "Gregory Bateson's 'Uncovery' of Ecological Aesthetics." *A Legacy for Living Systems*, edited by Jesper Hoffmeyer, vol. 2, Springer Netherlands, 2008, pp. 153–67. doi:10.1007/978-1-4020-6706-8_11.

Harrison, Helen Mayer, and Newton Harrison. *The Lagoon Cycle*. Herbert F Johnson Museum of Art, Cornell University, 1985.

Harrison, Helen Mayer, and Newton Harrison. *The Serpentine Lattice*. Douglas F. Cooley Memorial Art Gallery, Reed College, 1993.

Harrison, Helen Mayer, and Newton Harrison. *From There to Here*. The Harrison Studio, 2001.

Harrison, Helen Mayer, and Newton Harrison. *Peninsula Europe: The High Grounds*. The Harrison Studio, unpaginated, 2001.

Harrison, Helen Mayer, and Newton Harrison. "Knotted Ropes, Rings, Lattices and Lace: Retrofitting Biodiversity into the Cultural Landscape." *Biodiversity: A Challenge for Development, Research and Policy*, edited by Wilhelm Barthlott, Springer-Verlag, 2001.

Harrison, Helen Mayer, and Newton Harrison. "The Harrison Studio." *The Monongahela Conference on Post Industrial Community Development*. 23–25 Oct. 2003, Pittsburgh, PA. web.archive.org/web/20120517051232/http://moncon.greenmuseum.org/.

Harrison, Helen Mayer, and Newton Harrison. "Public Culture and Sustainable Practices: Peninsula Europe from an Ecodiversity Perspective, Posing Questions to Complexity Scientists." *Structure and Dynamics: Journal of Anthropological and Related Sciences*, vol. 2, no. 3, 2007, escholarship.org/uc/item/9hj3s753.

Harrison, Helen Mayer, and Newton Harrison. *The Time of the Force Majeure: After 45 Years Counterforce Is on the Horizon?* Prestel, 2016.

Harrison, Helen Mayer, and Newton Harrison. *On The Deep Wealth of this Nation, Scotland*. The Center for the Study of the Force Majeure, 2018.

Harvey, David. *Seventeen Contradictions and the End of Capitalism*. Profile, 2015.

Kagan, Sacha. *Art and Sustainability: Connecting Patterns for a Culture of Complexity*. 2nd ed., Transcript Verlag, 2013.

Klee, Paul. *Creative Confessions and Other Writings*. Tate, 2013.

Latour, Bruno. *Facing Gaia: Eight Lectures on the New Climactic Regime*. Polity, 2017.

McIntosh, Alastair. *Some Contributions of Geopoetics to Modern Scottish Land Consciousness*. Scottish Centre for Geopoetics, 29 Oct. 2017, www.geopoetics.org.uk/some-contributions-of-geopoetics-to-modern-scottish-land-consciousness-alastair-mcintosh/.

Missemer, Antoine. "Natural Capital as an Economic Concept, History and Contemporary Issues." *Ecological Economics*, vol. 143, Jan. 2018, pp. 90–96. *ScienceDirect*, doi:10.1016/j.ecolecon.2017.07.011.

O'Brien, Allyson, et al. "How Is Ecosystem Health Defined and Measured? A Critical Review of Freshwater and Estuarine Studies." *Ecological Indicators*, vol. 69, Oct. 2016, pp. 722–29. doi:10.1016/j.ecolind.2016.05.004.

Ozdemiroglu, Ece, and Rosie Hails. *Demystifying Economic Valuation*. 04, Valuing Nature Programme, 2016.

Rothenberg, Jerome. "Ethnopoetics at the Millennium: A Talk for the Modern Language Association." *Ubuweb Ethnopoetics: Discourses*. 29 Dec. 1994, www.ubu.com/ethno/discourses/rothenberg_millennium.html.

Roy, Arundhati. *Capitalism: A Ghost Story*. London: Verso Books, 2014.

Schön, Donald. "Generative Metaphor and Social Policy." *Metaphor and Thought*, edited by Andrew Ortony, Cambridge UP, 1993, pp. 138–63.

Shotter, John. "Goethe and the Refiguring of Intellectual Inquiry: From 'Aboutness'-Thinking to 'Withness'- Thinking in Everyday Life." *Janus Head*, vol. 8, no. 1, 2005, pp. 132–58.

Snyder, Gary. "The Politics of Ethnopoetics." *Ubuweb Ethnopoetics: Discourses*. www.ubu.com/ethno/discourses/snyder_politics.pdf. PDF file.

The Club of Rome. *The Predicament of Mankind. Quest for Structured Responses to Growing World-Wide Complexities and Uncertainities*, 1970, demosophia.com/wp-content/uploads/Predicament-Club-of-Rome-1970-1.pdf. PDF file.

Wilkins, Amy, editor. *Dancing Around the Bride: Cage, Cunningham, Johns, Rauschenberg and Duchamp*. Philadelphia Museum of Art, 2013.

CHAPTER 6

New Worlds Beyond Reality: Imagined Futures in Laura Gallego's *Las hijas de Tara*

Irene Sanz-Alonso

Abstract

The purpose of this work is to explore how literary genres such as fantasy and science fiction, without the time and place constraints of realistic types of writing, enable authors to imagine alternative worlds based on sustainability and a respectful relationship with the other. This chapter will analyse Laura Gallego's *Las hijas de Tara* (*Tara's Daughters*) following ecocritical and ecofeminist theories in order to establish a contrast with the unhealthy social systems that exist in most of our world. In first place, the portrayal of the planet Tara will be explored following Lovelock's hypothesis of Gaia, and the positive and negative implications of presenting a planet as a living entity will be assessed. The analysis will then focus on the values, practices and attitudes that make of the societies portrayed in the novel an example (or not) of healthy co-habitation of individuals with their surroundings, and in so doing concepts such as the interconnectedness of life will be dealt with, together with the dichotomy natural/artificial. Therefore, this chapter aims at showing how fantasy and science fiction works may help readers imagine alternative, and more sustainable, futures.

Keywords

Fantasy, Laura Gallego, *Las hijas de Tara,* nature vs technology.

1 Introduction

Literature has always been a powerful resource to imagine alternative worlds, and genres such as science fiction and fantasy enable writers to do so without the constraints imposed by reality. Using some of the traditional imagery of such literary genres—as robots or magic—in *Las hijas de Tara* [*Tara's Daughters*] the Spanish writer Laura Gallego portrays a future Earth where nature and technology are at war, and two human civilizations with opposing lifestyles live

© KONINKLIJKE BRILL NV, LEIDEN, 2022 | DOI:10.1163/9789004501270_008

confronted. In the following pages Gallego's novel will be explored focusing on the relation both civilizations have with the environment. One of the main aspects of this analysis is how Gallego's imagined future, despite pollution and violence, harbours hope in terms of sustainability and ecological awareness.

Laura Gallego is one of Spain's most acclaimed writers of young adult and fantasy fiction. She started publishing her works at an early age when she won the 'Barco de Vapor' Prize in 1999 for her work *Finis Mundi*. She is especially well-known for her trilogy *Memorias de Idhún*; and in 2012 won the National Prize of Children and Young Adult Literature for her novel *Donde los árboles cantan*, in which the natural world plays a very important role—as in *Las hijas de Tara*. The novel was published in 2002, although originally it was not conceived as a book but as a script for an animated film in 3D which was never developed (Gallego, *Laura Gallego* n.p.). The novel has never been translated into other languages, although other works by Gallego have, and in February 2018 it was republished by the Spanish publishing house Minotauro, which specializes in science fiction and fantasy.

2 Gallego's Mother Earth: Tara

In *Las hijas de Tara* Gallego sets the story on a very different Earth to the one we live in nowadays since wars and pollution have left its surface divided into two main areas: the wild natural region in constant expansion known as Mannawinard; and the cities, both static and nomadic, which are called *dumas* and are characterized by a strong presence of artificial intelligences. Between the two areas there are vast dangerous regions known as the Wastelands, deserted territories poisoned by pollution and radiation. Just as the Earth is geographically polarized, so is humanity. Most humans, known as urbanites, live in the dumas. On the other hand, the civilization that supports Mannawinard is divided into three main groups: Tara's Daughters, who are similar to a female religious organization; the Ruadh, warriors that live in the frontier with the dumas; and the wildlings that inhabit the natural world. The author describes Mannawinard in terms of wildness and lushness, as an uncontrollable entity with a personality of its own. The first time Mannawinard, or nature, appears in the story is when an urbanite named Kurt, who is lethally infected by radiation, comes across some of Tara's Daughters. Since he has grown up in a duma, he has never had contact with nature until that moment, but he recalls what he was told when he was young: "[…] he had never seen anything like that. But he had heard about it, he had been explained everything about it in his history classes at school, since he was a child. It was called Nature. And it was the enemy of men and their technology" (11).[1]

To understand the origin of the conflict between nature and technology in the novel we need to consider the Song of Mannawinard, a set of lines similar to a prayer that Tara's Daughters recite when they want to meditate or to recover strength:

> It was in the old days,
> before Mannawinard,
> when men, arrogant,
> destroyed the earth [...].
> And the goddess Tara cried, cried,
> but men were deaf,
> To their mother's voice.
> And the goddess Tara unleashed her fury,
> and she was Mannawinard. (37; emphasis in original)[2]

The song continues describing how when nature started to devour civilization and technology a small group of humans decided not to take revenge on Mannawinard but instead to change their lifestyle and embrace Tara: "... some of the survivors had renounced the remains of their technological world and their artificial way of life in order to go back to the jungle and to live under the auspice of the Goddess Mother, in harmony with nature, respecting the balance of Mannawinard" (38).[3] The novel shows that the partial destruction of humanity at the hands of nature—also called Mannawinard, "the evergreen land of the goddess Tara" (35)[4]—was a direct consequence of humans' actions on the environment through pollution and radiation, so Tara evolved into Mannawinard as a kind of revenge.

---

1 "... nunca había visto nada semejante. Pero había oído hablar de ello, le habían explicado todo aquello en las clases de historia del colegio, desde que era un niño. Se llamaba Naturaleza. Y era enemiga del hombre y su tecnología". All the translations of fragments from the novel have been done by the author.

2 *Fue en los días antiguos, / antes de Mannawinard, / cuando el hombre, soberbio, / destruía la tierra* [...]. */ Y la diosa Tara lloraba, lloraba, / pero los hombres eran sordos, / a su voz de madre. / Y la diosa Tara desató su furia, / y fue Mannawinard. /* (37)

3 "... algunos de los supervivientes habían renunciado a los restos de su mundo tecnológico y a su modo de vida artificial para volver a la selva y vivir en el seno de la Diosa Madre, en armonía con la naturaleza, respetando el equilibrio de Mannawinard".

4 "la tierra siempre verde de la diosa Tara".

In the novel the natural world is referred to as Mannawinard, and it is portrayed as the representation of the goddess Tara, who in turn is part of every living being. The association of women and nature has been subjected to much analysis, especially in the last decades with the emergence of ecofeminism, which focuses on the relationship between women and nature in terms of oppression of the *other* and on how that oppression can be eradicated. This association has always existed, and its two main representations are accurately portrayed in the novel. On the one hand, since Prehistoric times nature has been described in feminine terms as "a kindly beneficent female who provided for the needs of mankind in an ordered, planned universe" (Merchant, *Earthcare* 77). On the other hand, there are several examples in which nature has also been represented as a temperamental woman characterized by storms, hurricanes, droughts, chaos and other natural disasters (Merchant, *Earthcare* 77; Li 277). In the novel Mannawinard is described as both a nurturing presence that gives life, but at the same time—and perhaps with more emphasis because of the revenge component—as a cruel entity that destroys anything artificial in its path, as Tara's weapon against pollution and environmental degradation.

Gallego's Tara echoes the organic worldview of prehistoric societies, which worshipped Mother Earth as the goddess and creator of life. Cultural ecofeminism, as Eric C. Otto points out, fosters this "reemergence of ancient matriarchal belief" because they were characterized by "peace and respect for all life" (17). In her analysis of the association between nature and women, Carolyn Merchant highlights that in these ancient civilizations nature was seen as a nurturing mother, "sensitive, alive, and responsive to human action" ("Mining" 102). Since the planet, and all its inhabitants, were seen as part of a whole, the Earth was not conceived as "an object for exploitation and domination" (Eisler 23) since "one does not readily slay a mother, dig into her entrails for gold, or mutilate her body" (Merchant, "Mining" 100). However, this vision of nature changed with the scientific revolution when nature became an endless resource to be exploited and mastered, thus losing its sacred character and its organic status. The concept of the Earth as a female entity called Gaia has generated much debate both among feminist and among environmentalists. Even though the concept of Gaia has a long history, in the 20th century it re-emerged thanks to the hypothesis presented by scientist James Lovelock in his book *Gaia: A New Look at Life on Earth* (1979). According to this hypothesis, the Earth behaves as an organism and all life is interconnected in such a way "that the mean global temperature, the composition of reactive gases in the atmosphere, and the salinity and alkalinity of the oceans are not only influenced but regulated, at a planetary level, by the flora, fauna, and microorganisms"

(Sagan and Margulis 353). Lovelock's idea of a planet and all its lifeforms conceived as an interconnected web represents "a scientific update of the belief system of Goddess-worshipping prehistoric societies" (Eisler 26). Even though ecologists and feminists reacted positively to the hypothesis at first, because it gave nature an active role once more, with the passing of time the hypothesis became highly criticized by environmentalists. One of the most controversial points of Lovelock's statements was that the Earth was conceived as a set of "self-regulating mechanisms" (Kheel 251), and as such, the planet was able to recover from damaging human practices. According to this view, if nature is able to regulate itself then humans are free to exploit and pollute it because no matter what damage is done, the Earth will find a mechanism to reverse it. In *Literature, Nature and Other* Patrick D. Murphy also criticizes the concept of a self-healing Earth because if that were scientifically true, then there would be no need for environmentally-friendly attitudes (24).

Another controversy surrounding the hypothesis is the very name Lovelock chose. Gaia makes reference to the Greek Earth goddess and by choosing that name the scientist somehow gave the planet a feminine character, reinforced by the essentialist idea of nature and women as linked because of their ability to create life. Murphy comments on the problems of the choice of the name in a patriarchal context because "Gaia designates a female entity; designating an entity female in a patriarchal culture guarantees its subservient status" (61). Other authors have also commented on the problematic implications of choosing such a name. For example, Greta Gaard posits that the concept of Gaia, as that of Mother Earth, "dooms nature to a female way of being, which in Western culture means a subordinate way of being" (305). Similarly, Catherine Roach highlights the negative connotations of the idea of Mother Earth: "engendering the Earth as female mother" in a patriarchal culture "will not achieve the desired aim of making our behavior more environmentally sound, but will instead help to maintain the mutually supportive, exploitative stances we take toward our mothers and toward our environment" (52). Therefore, as seen above, referring to the Earth as a female entity can be counterproductive because instead of recovering the mythical and more pro-ecological view of nature and of the planet as a goddess-like organism, in patriarchal systems of belief referring to the planet as feminine involves its passivity.

In the novel, Gallego rewrites the myth of Mother Earth and she creates her own version of the Gaia hypothesis. The planet is referred to as an interconnected web of life known as Tara. She also refers to Tara as Goddess Mother, but despite its feminine nature, the planet is not described in terms of passivity and exploitation but rather as a vengeful and active organism. In contrast with some of the interpretations of Lovelock's Gaia hypothesis, Tara is not seen as an organism that regulates itself when pollution and environmental degradation take place, but rather as an active system that does not only regulate itself but that also punishes the ones who damage it. An example of this can be seen when the advance of Mannawinard takes place in the novel and those who support Tara justify her destructive ways: "Because humans have forgotten that sacred link that connects us all with Mother Earth. And she is just defending herself" (109).[5]

Mannawinard, which is described as Tara's executing arm, is a growing jungle that spreads whenever Tara is attacked, as a kind of revenge. The most interesting aspect of Mannawinard is that despite its destructive character, especially in the eyes of urbanites because it tears down dumas and kills their inhabitants, it is a source of life in all its forms:

> And the cause of that catastrophe was something alive. The green line of the horizon was the limit of Mannawinard, the horrific killer jungle. Trees had sprouted from the ground, growing at an awful speed, destroying the streets of the city. Huge vines had climbed the tall buildings of steel and glass, wrapping them with their lethal embrace, smothering them until making them burst. (122)[6]

Mannawinard appears portrayed as a voracious jungle by urbanites, but even those who believe in Tara are aware of how powerful and dangerous it is despite its beauty. For example, Keyko, who belongs to Tara's Daughters and is able to listen to Tara's voice and use her magic, is fascinated and at the same time a little frightened when she contemplates Mannawinard in its splendour: "The jungle did not seem the same in that darkness, and for a moment she understood why urban-dwellers were so afraid of Mannawinard. In the land of the Goddess Mother, human beings were again a simple link in the chain of life: they could eat or be eaten" (148).[7] Therefore, Gallego's Gaia or Tara is seen as web of life but not as a passive organism that simply waits to be damaged or exploited. In this novel the Earth is portrayed as a living creature that counter

5 "Porque los hombres han olvidado ese sagrado vínculo que nos une a todos con la Madre Tierra. Y ella solo se defiende."

6 "Y el causante de esa catástrofe era algo vivo. La línea verde del horizonte era el límite de Mannawinard, la espantosa selva asesina. Los árboles habían brotado del suelo, creciendo a una velocidad endiablada, destruyendo las calles de la ciudad. Enormes enredaderas habían trepado por los altos edificios de acero y cristal, envolviéndolos con su abrazo letal, asfixiándolos hasta hacerlos reventar."

7 "La selva no parecía la misma en aquella oscuridad, y por un momento comprendió por qué los urbanitas temían tanto a Mannawinard. En la tierra de la Diosa Madre, el ser humano volvía a ser un simple eslabón en la cadena de la vida: podía comer, o ser comido".

attacks when damaged. She responds to artificiality and weapons with the proliferation of non-human animals and lush vegetation:

> Mannawinard was much more than a forest or a jungle. It was much more than a natural world. It was, to say it in some way, an explosion of life. Trees, flowers, animals… all was gigantic, as if a huge vital energy impregnated everything. The catalogue of living species had surprisingly multiplied in comparison with the old days; ancient biology had been overwhelmed with such a flood of new species. It was as if nature had wanted to challenge itself and show the world all the wonders it was capable of creating.
>
> Mannawinard was much more than a forest or a jungle, the wildlings knew it. Mannawinard was the rebirth of the Goddess Mother. (171)[8]

Mannawinard, which represents Tara and her power, has a destructive component, which results ambiguous because that destruction is carried out through an explosion of life, and once Mannawinard conquers a duma what is left behind is a wild forest populated by numerous creatures. The relationship with Tara and Mannawinard is what divides humans into two opposing bands, although Tara is presented as always willing to give an opportunity to those who embrace her and are able to hear her voice, as it happens with Kim.

Another interesting aspect regarding Tara is the spirituality that surrounds her. In the novel she is referred to as a web of life but also as a kind of deity, and Kim, the urbanite that is one of the protagonists of the story, criticises this by stating that, by worshiping a goddess, Keyko's life is limited since she has to behave according to certain dogmas. Keyko rejects the criticism claiming that Tara's Daughters do not pray, that the only things they had to do regarding Tara were "to believe in her, to respect the balance of Mannawinard, and to be part of the wheel of life" (39).[9] Keyko continues saying that Tara is not a goddess to be worshiped: "Tara is the Earth. Tara is us, all of us. Protecting Tara we protect

8 "Mannawinard era mucho más que un bosque o una selva. Era mucho más que un mundo natural. Era, por decirlo de algún modo, una explosión de vida. Los árboles, las flores, los animales… todo era gigantesco, como si una gran energía vital lo impregnase todo. El catálogo de especies vivas se había multiplicado sorprendentemente con respecto a los días antiguos; la biología antigua se habría visto desbordada ante semejante aluvión de nuevas especies. Era como si la naturaleza hubiera querido retarse a sí misma y mostrar al mundo todas las maravillas que era capaz de crear.

Mannawinard era mucho más que un bosque o una selva, los salvajes lo sabían. Mannawinard era el renacimiento de la Diosa Madre."

9 "creer en ella, respetar el equilibrio de Mannawinard, formar parte de la rueda de la vida".

ourselves. You urbanites do not want to admit that if Mannawinard had not appeared, human beings would have destroyed themselves a long time ago" (65).[10] The idea that Tara is the Earth is restated several times throughout the novel, reinforcing the idea to the extent that some readers may find it a bit repetitive. Nevertheless, this over-didacticism could respond to the fact that the targeted readership is mainly children and young adults, which is reflected in the narrative style of the book.

3 Natural vs Artificial

A relevant element in the novel in relation the idea of the interconnectedness of life is the portrayal of the relationship between humans and non-human animals and plants. As we have already seen, those who support Tara and are able to hear her voice see the Earth as an interconnected web of life, so plants and non-human animals are also part of this web: "We are all part of the world, child. We are the world. If we harm Tara, we harm ourselves. Never forget this" (219).[11] In this sense, non-human animals and plants are seen as equals, as another link in the chain of life that Tara represents. Keyko had little experience with non-human animals while she lived with Tara's Daughters because their temple was near the deserted Wastelands, but her knowledge of different species through study was enough to be able to appreciate natural creation and not be afraid of them because of ignorance. On the other hand, her companion on the journey, Kim, who had always lived in dumas, had no experience whatsoever with non-human animals: "She had never seen an animal so big. And, to tell the truth, before entering the Wastelands she had never seen an animal" (51).[12] When they arrive at Mannawinard, which has just conquered one of the largest dumas in the frontier, Duma Murias, both Keyko and Kim look perplexed at the explosion of life that is conquering the city through animals and vegetation. The author decides to narrate the arrival through Kim, which is quite significant because of her upbringing as an urbanite she despises anything natural: "And then Kim heard the sounds. Inhuman shrieks, shrieks from strange animals that had invaded the city and had turned it into

10 "Tara es la Tierra. Tara somos nosotros, todos nosotros. Protegiendo a Tara, nos protegemos a nosotros mismos. Los urbanitas no queréis reconocer que si no hubiese aparecido Mannawinard el ser humano ya haría tiempo que se habría destruido a sí mismo".

11 "Todos somos parte del mundo, hija. Nosotros somos el mundo. Si dañamos a Tara, nos dañamos a nosotros mismos".

12 "Nunca había visto un animal tan grande. Y, para hacer honor a la verdad, hasta que se había internado en los Páramos nunca había visto un animal".

their new home. She had never heard the sounds of the jungle, and found them simply frightening" (122).[13] While they stay in what used to be Duma Murias, Kim is continuously suspicious of everything that surrounds her, even if there is no direct threat:

> From the highest places, whether trees or half-fallen buildings, a great variety of birds with colourful feathers emitted peculiar sounds jumping from branch to branch or flying over the heads of the new arrival, something that worried Kim. From time to time she discovered some hairy creature looking directly at her from the top of a tree, and she had to control herself not to grab her weapon and fire. (124–25)[14]

As we can see from Kim's perspective, anything which is not human—or even urbanite—or artificial disturbs her. She reacts to organic creatures with fright, worry and even disgust:

> But the worst thing were the beings that walked on the floor. There were small rodents, reptiles, mammals that ran to hide when they saw her, and Kim was startled in shock. Once she put her hand on a cracked wall and she was about to graze a gigantic black and hairy bug with many legs, truly disgusting. (125)[15]

When Kim thinks about non-human animals and their absence in the dumas, she recalls the use of little company robots as pets, made by one of the companies that controls the dumas with their technology. The reason for choosing artificial creatures as pets and not real animals was that "'real' animals were unpredictable and annoying, and for centuries they had been considered a

---

13 "Y entonces, Kim oyó los sonidos. Gritos inhumanos, gritos que procedían de extraños animales que habían invadido la ciudad y habían hecho de ella su nuevo hogar. Ella nunca había escuchado los sonidos de la selva, y le parecieron sencillamente aterradores".

14 "Desde los lugares más altos, ya fueran árboles o edificios semiderruidos, una gran variedad de aves de plumaje colorido emitía peculiares sonidos, saltando de rama en rama o sobrevolando las cabezas de los recién llegados, cosa que inquietaba mucho a Kim. De vez en cuando descubría a alguna criatura peluda mirándola fijamente desde lo alto de un árbol, y tenía que controlarse para no sacar su arma y disparar".

15 "Pero lo peor eran los seres que caminaban por el suelo. Había pequeños roedores, reptiles, mamíferos que corrían a ocultarse a su paso, y Kim iba de sobresalto en sobresalto. Una vez apoyó la mano en una pared agrietada y estuvo a punto de rozar un gigantesco bicho negro y peludo con muchas patas, verdaderamente repugnante".

threat to the system" (125).[16] The lack of non-human animals in the dumas is also reflected in the food industry since some companies had been able to create an artificial creature with all the proteins, vitamins and minerals needed by the human body to develop healthily, and so this creature was cloned once and again and distributed across the dumas. It is interesting to note that urbanites see this kind of artificial diet as something positive because it provides all the necessary nutrients without having to resort to nature:

> Kim would have felt unable to hunt animals in the jungle, as wildlings did, and then eat them without knowing what she was having, without knowing if that creature suffered from some disease… she shivered involuntarily, just by thinking about it. The very idea seemed disgusting. (125)[17]

Urbanites do not need to kill animals to have a complete diet because they already have the technological advances that allow them to recreate natural substances present in non-human animals and necessary for the development of the human body. This issue could be interesting from a vegetarian and vegan point of view, but the reason why urbanites in the novel reject products coming from non-human animals has nothing to do with the ethical implications or the infrastructure of food industries; they prefer artificial nutrients because they refuse anything that comes from nature. On the other hand, it is significant that wildlings do eat animal products despite their respect for non-human species. For this reason, when they hunt other species, they do so without using painful methods because that would go against their set of beliefs. In other words, when wildlings go hunting they do so considering they are both predators and prey, and that they confront other creatures face to face.

Wildlings' respectful attitude towards nature is not only seen in how they treat non-human animals but also in how they manage natural resources such as plants and trees. For example, they explain that trees are only cut when they need to build houses and not for some other minor things because "After all, they are also children of Tara, they have not hurt us and they cannot defend themselves" (226).[18] In the case of urbanites, they are not able to appreci-

16 "Los animales 'de verdad' eran imprevisibles y molestos, y hacía siglos que eran considerados una amenaza al sistema".

17 "Kim se habría sentido incapaz de cazar animales en la selva, como hacían los salvajes, y luego comérselos sin saber lo que comía, sin saber si aquel bicho padecía alguna enfermedad… Se estremeció involuntariamente, solo de pensarlo. La simple idea le resultaba repugnante".

18 "Al fin y al cabo, ellos también son hijos de Tara, no nos han hecho daño y no pueden defenderse".

ate vegetation since in the dumas everything natural is destroyed. However, dumas do have artificial trees that imitate real ones: "There was no need to water them, they did not dry, they did not need soil, they did not produce pollen" (125–26).[19] Apart from the artificial trees in the dumas, other types of vegetation had also been replaced by artificial counterparts, just as non-human animals: "Vegetal proteins had not been completely removed from the diet, anyway; Tong-Pao was able to synthesize them artificially, and it introduced them in some of its products" (126).[20] Once more we can see how in their effort to get rid of anything coming from nature, they use their scientific knowledge to imitate nature through technology and genetics.

Even though the lifestyles in Mannawinard and the dumas are not explained in detail—we only get some glimpses through the protagonists' experiences and memories—we can highlight that urbanites live in a futuristic Earth in which artificiality is the norm, whereas those who embrace Tara live as humans lived in the ancient days, and in harmony with nature. Because of their particular ways and customs, urbanites think that the inhabitants of Mannawinard "live like animals" (84).[21] When Kim, the urbanite protagonist of the story, describes how Mannawinard conquered the Earth once humans' aggressive ways towards Tara had reached a certain point, she does so by despising anything coming from nature. For example, she talks about "murderous plants", and about how "nature turned against humans and literally smothered all their cities" (84).[22] Kim's words illustrate an overtly anthropocentric attitude seeing the planet as a place to be conquered by humans, and without being able to understand that Tara's aggressive response was in proportion to the damage caused.

We can also see that as a result of this anthropocentric thinking urbanites hyperseparate themselves from nature, and thus see everything natural as negative: "'Nature is everything which has not been created by the hand of man,' Kim replied. 'Nature is chaos, disorder, turmoil, danger, threat'" (84).[23] Kim's anthropocentric statement is refuted by Keyko's words—which echo Lovelock's concept of Gaia—on how difficult it is for human beings to understand that they are not the centre of the world: "Nature has its own balance,

19 "No hacía falta regarlos, no se secaban, no necesitaban tierra, no producían polen".

20 "Las proteínas vegetales no habían sido totalmente desterradas de la dieta, de todas formas; la Tong-Pao era capaz de sintetizarlas artificialmente, y las introducía en algunos de sus productos".

21 "…vive como animales".

22 "La naturaleza se volvió contra el hombre y ahogó literalmente todas sus ciudades".

23 "Naturaleza es todo lo que no ha creado la mano del hombre -replicó Kim-. Naturaleza es caos, desorden, descontrol, peligro, amenaza".

what happens is that you, urbanites, think it is annoying to have something over human beings, and you do not admit that regarding creation Tara is well ahead of you" (84).[24] The ability to create—or imitate life—is precisely one of the most significant aspects of the novel since that is what urbanites try to do as a challenge to nature. When men tried to conquer nature, Mannawinard appeared as Tara's weapon to counterattack, and at that point humans lost control over nature. Since then, humans have been trying to "recreate" nature through artificial means because that is the only way they can control it and use it for their own benefit. By controlling the power of creation, even if it is artificial, urbanites have placed themselves as the masters of their world.

Apart from the corporations, those who control the power in the dumas are the Ideologists of Progress. They are described as ancient human beings that lived before Mannawinard existed, and whose brains have been transplanted into artificial bodies. Although they are not worshiped as gods, they are portrayed as powerful beings that manipulate everyone living in the dumas. In order to keep their power they invented the "Myth of the Artificial", which states that:

> Human beings are owners and masters of the Earth. We have the right to change it and create it again. Our reason can control everything. Nothing is more valuable than what we have created. Everything natural is wrong. Nature is not subjected to us: we must control it. Wildlings have resurrected magic, and they attack us with it: we must destroy them because they threaten our technological civilization. (169)[25]

Even though the message of the "Myth of the Artificial" is quite simplistic—and to explain this we need to appeal again to the intended readership of the book—it is clear in its portrayal of the exploitation of the Earth at the hands of humans, and as such, urbanites are presented as egocentric and speciesist. Besides, this statement does not only justify the annihilation of anything natural but also the murder of wildlings, who represent a real threat.

24 "La naturaleza tiene su propio equilibrio, lo que pasa es que a vosotros, los urbanitas, os molesta la idea de que haya algo superior al ser humano, y no reconocéis que en materia de creación Tara os ha aventajado con creces".

25 "El hombre es el dueño y señor de la Tierra. Tenemos derecho a cambiarla y crearla otra vez. Nuestra razón puede controlarlo todo. Nada vale más que lo hemos creado nosotros. Todo lo natural es malo. La naturaleza no se somete a nosotros: debemos controlarla. Los salvajes han resucitado la magia, y nos atacan con ella: debemos destruirlos, porque ponen en peligro nuestra civilización tecnológica".

Urbanites see wildlings as inferior beings, in a similar position to that of non-human animals, because of their way of life and their repudiation of artificiality. However, those who control the dumas are aware of the danger that wildlings' knowledge of magic entails. Wildlings also despise urbanites because they reject Mother Earth and insist on manipulating creation through technology. In the novel the two sides are represented by Kim and Keyko, and although Keyko is open-minded and tries to show Kim the benefits of Tara and of magic, the urbanite is suspicious of her companion most of the time, even after Keyko saves her life. We can also perceive how much most wildlings hate urbanites; for example, some Ruadh do not doubt in attacking and killing urbanites because of their status as urbanites: "Since I was a child I had been taught not to trust urbanites nor the dumas, and to be suspicious of anything coming from them… especially of those who carry weapons and of artificial beings" (54).[26]

Even though the hatred between urbanites and wildlings can be perceived at the beginning of the relationship between Keyko and Kim, we also see how little by little they are able to overcome their differences and understand that they are more similar than they thought. In the middle of this tense relationship is the biobot AD-23674-M—significantly renamed as Adam—which is stolen by Kim on a secret mission. Throughout the novel Kim and Keyko discover that the biobot is not a normal robot that can be programmed, and at the end they are told that two powerful druids had been able to infuse the biobot with a soul, and that is why Kim has to escape and needs Keyko's help. The biobot functions as a kind of nexus between the two humans because it shows Keyko that artificial life is not always dangerous and wrong; while it teaches Kim to appreciate her life and what she can learn from others. In a world which is divided between nature and technology, the biobot trespasses all boundaries and becomes a central element in the story, a bridge between Mannawinard and the urbanites.

4 Conclusion

Therefore, *Las hijas de Tara* is a novel about a journey, both physical and psychological, that implies learning from the other in order to make the world a better place. Despite having been told since she was a child that anything artificial was wrong, Keyko does not hesitate to help a technologically-modified

26 "Desde niña me han enseñado a desconfiar de los urbanitas y de las dumas, y de todo lo que provenga de ellas… muy especialmente de aquellos que portan armas y de los seres artificiales".

Kim when they meet for the first time. Kim's reluctance little by little evolves into a desire of learning more about Mannawinard despite her prejudices. At the end of the novel Kim is even capable of hearing Tara's voice and of learning from Mannawinard: "She placed herself on the branch and waited in silence, as she had seen the panthers and other big felines do in Mannawinard" (228).[27] Through her journey Kim is able to see beyond her unfounded hatred and appreciate other ways of life based on respect towards the natural world. When she hears Tara's voice, she is able to understand what Keyko had told her repeatedly: "Just remember that everything is just a single thing, that, from the smallest insect of Mannawinard to the last urbanite in the dumas, we are all part of the same world, we are all part of the Earth, of Mother Tara" (218).[28]

By providing the planet with a voice of its own Laura Gallego gives visibility to the exploitation and degradation to which the Earth has been—and still is—subjected. Although her novel is set in a distant future in which technology and nature have become what seem to be irreconcilable enemies, the problematic around the environment reflects much of today's current situation. Despite the repetition of some of the ideas around Tara and a somewhat simplistic narrative style, the novel makes readers reflect on the relationship between human beings and the other creatures with which we share the planet, as well as on how we relate with the planet itself. Tara makes us see the Earth as an entity which is able to suffer but which is also able to answer back, an idea similar to Lovelock's Gaia and her ability to regain balance when lost. Gallego's Tara teaches readers the lesson that humans are just a link in the chain of life and that we are responsible for our actions towards others. It also shows us that human beings need to reflect on their role within this chain and on how important it is to keep the balance of the Earth. Reading the novel as an adult allows room to reflect on the possible responses of children and young adult readers, and it is hopeful to think that this kind of literature—both entertaining and with an ecological message—can show young readers that there are different ways of relating with nature and that they have the possibility of choosing which one they prefer. Science fiction books like *Las hijas de Tara* allow us as adults to imagine a more sustainable world for children and give us the tools to make it possible by putting into practice—at least literarily—what ecocritics and ecofeminists expose in their theoretical works.

---

27 "Se acomodó sobre la rama y aguardó en silencio, como había visto hacer a las panteras y otros grandes felinos de Mannawinard".

28 "Recuerda que todo es una sola cosa, que, desde el más pequeño insecto de Mannawinard hasta el último urbanita de las dumas, todos formamos parte de un mismo mundo, todos formamos parte de la Tierra, de la Madre Tara".

Works Cited

Eisler, Riane. "The Gaia Tradition and the Partnership Future: an Ecofeminist Manifesto." *Reweaving the* World, edited by Irene Diamond and Gloria F. Orenstein, Sierra Club, 1990, pp. 23–34.

Gaard, Greta. "Ecofeminism and Native American Cultures: Pushing the Limits of Cultural Imperialism." *Ecofeminism. Women, Animals, Nature*, edited by Greta Gaard, Temple UP, 1993, pp. 295–314.

Gallego García, Laura. *Las hijas de Tara.* 2014. Ediciones SM, 2002.

Gallego García, Laura. *Laura Gallego.* www.lauragallego.com/libros/novelas-independientes/las-hijas-de-tara/. Accessed 24 Feb. 2018.

Kheel, Marti. "From Heroic to Holistic Ethics: The Ecofeminist Challenge." *Ecofeminism. Women, Animals,* Nature, edited by Greta Gaard, Temple UP, 1993, pp. 243–71.

Li, Huey-Li. "A Cross- Cultural Critique of Ecofeminism." *Ecofeminism. Women, Animals, Nature*, edited by Greta Gaard, Temple UP, 1993, pp. 272–94.

Lovelock, James. *Gaia: A New Look at life on Earth.* 1979. OUP, 2016.

Merchant, Carolyn. *Earthcare. Women and the Environment.* Routledge, 1996.

Merchant, Carolyn. "Mining the Earth's Womb." *Machina Ex Dea,* edited by Joan Rothschild, Pergamon, 1983, pp. 99–117.

Murphy, Patrick D. *Literature, Nature, and Other. Ecofeminist Critiques.* State U of New York P, 1995.

Otto, Eric C. "Ecofeminist Theories of Liberation in the Science Fiction of Sally Miller Gearhart, Ursula K. Le Guin, and Joan Slonczewski." *Feminist Ecocriticism, Environment, Women, and Literature*, edited by Douglas A. Vakoch, Lexington, 2012, pp. 13–37.

Roach, Catherine. "Loving Your Mother: On the Woman-Nature Relation." *Ecological Feminist Philosophies*, edited by Karen J. Warren, Indiana UP, 1996, pp. 52–65.

Sagan, Dorion, and Lynn Margulis. "God, Gaia and Biophilia." *The Biophilia Hypothesis*, edited by Stephen R. Keller and Edward O Wilson, Washington, 1993, pp. 345–64.

CHAPTER 7

Simon Ortiz's Narrative and Joy Harjo's Poems: Towards Regenerative Societies and New Worlds

Imelda Martín-Junquera

Abstract

Simon Ortiz's poetry mainly focuses on the recovery of a "storytelling" tradition and the culture of the indigenous American ancestry at large. His stories, however, tend to explore different themes and so, they appear as windows which open to new possibilities for the improvement of the life conditions for Native Americans. Harjo, on her part, deals with everyday life events and how they acquire significance for her. At the same time, the natural and human element appear in combination, in perfect symbiosis with this traditional culture in both writers who struggle with their literary productions to produce a significant change towards the creation of interspecies relationships in their local areas. *Men on the Moon* by Ortiz and the narrative poems by Joy Harjo included in *A Map to the Next World: Poems and Tales* represent an escape from the isolation and marginalization of the human being, an exit towards a transformation which projects a hope and a transition to contemplate the creation of regenerative communities in ecocritical and posthuman terms.

Keywords

Posthuman, human, more-than-human, interspecies dialogue, Native American tribes.

1 Ortiz and Harjo: Heritage and Common Features

Simon Ortiz and Joy Harjo share not only their Native American origins but also the same preoccupations in their poetry and narrative. I am going to explore here texts both poems and short stories having to do with the transformation of the land and the world they inhabit at large; they partake, as well in their experience of language related trauma. Ortiz, having been forced to abandon his tribal language and learn English at school where he even had to deal with punishment inflicted on him when resorting to a language other than English,

© KONINKLIJKE BRILL NV, LEIDEN, 2022 | DOI:10.1163/9789004501270_009

while Harjo, who suffers her inability to express herself in a language different from English, makes conscious efforts to introduce traces of oral tradition in her writing.

I have chosen several short stories from *Men on the Moon: Collected Short Stories* ("Men on the Moon," "The End of Old Horse," "Feathers," "To Change Life in a Good Way," "What Indians Do" and "Distance") by Ortiz and some tales from *A Map to the Next World: Poems and Tales* ("A Map to the Next World," "the crow and the snake,"[1] "twins meet up with monsters in the glittering city," "when my son was born," "earthly desire" and "humans aren't the only makers of poetry" by Joy Harjo because they highlight the these authors's innovative cosmovision of the environment since they include not only positive relationships between the human and the more-than-human world which they share, but also their immense hope of the potential of the new generations to restore a sense of dignity to nature and its inhabitants.

The influence of the place where they were born and raised can easily be felt both in their poetry and prose. The concepts of home and place attachment, in connection with the problems related to the land at large, are widely explored in their works. Consequently, they have a strong connection with New Mexico; Ortiz, born in Albuquerque, belongs to the Acoma pueblo and started a degree at the University of New Mexico, although he finally graduated at the University of Iowa. In the case of Harjo, she is a member of the Mvskoke Nation and was born in Tulsa, Oklahoma but she finally studied at the University of New Mexico; thus, the landscape of the area, populated by the mythical juniper tree, the Sandia Mountains and the Rio Grande, among other emblematic landmarks, fill their writings. Another common feature that must be emphasized in the narrative and poetry of Ortiz and Harjo has to do with the concern they express about the future of the Southwest, in especial the care and protection of the surrounding landscape, without disregarding that it is impossible for the human race not to inflict inevitable transformations and to alter it somehow, as well as keeping alive the links they maintain with tradition as I have already pointed out in "The Inscription of the American Southwest in Navajo Tribal Parks" (2017). In the two poems analysed in that essay, Ortiz reflects on the landscape; he feels an intimate relationship with it, sitting on rocks for hours and listening to the message of the history inscribed on them; thus, he doesn't contemplate it as a romanticisation of the wilderness but rather with nostalgia, remembering how his presence and that of his people have been excluded from those now protected national parks.

1 Whenever titles by Harjo's stories are not capitalized it is because of the author's choice.

The influence of the oral tradition can be felt in their works as Donelle Dreese argues: "For Harjo, descriptions of landscapes and settings from the oral tradition are combined with modern cityscapes, often creating a surrealistic and sometimes disjunctive plane of existence" (38). Harjo herself finds difficulties in adjusting to the classifications of literary genres and, thus, she states in the introduction to *How we Became Human* "There is no separation between poetry, the stories and events that link them, or the music that holds all together, just as there is no separation between human, animal, plant, sky and earth" (loc. 270). In her writings, the common denominator seems to be the attempt to define the type of relationships that can be established between the human species and at least another animal or vegetal species and how to make them successful not only in terms of the profit that can be obtained by humans but, so that, they become significant for non-human others as well. A critique of capitalism and consumerism also permeates the pages, since these writers establish a compromise with their surroundings that takes them to consider and explore the use and overuse of natural resources as well as the importance of a relationship of interdependence among the species inhabiting a given ecosystem, or "trans species solidarity" (67) in the words of Braidotti. As Louise Westling argues in "Literature, Environment and the Posthuman," the big challenge in modern fiction consists of reflecting the intertwining of humans within a living community of many voices and being aware of their sounds, and distinguishing their different rhythms as well as the importance of non-human forces of nature. Criticising speciesism and also exploring the relationships between human beings and the more than human world, Rosi Braidotti provides a theory of decentring anthropocentrism which she defends as different from posthumanism because it adds a more positive turn to the nature/culture dualism:

> the posthuman dimension of anthropocentrism can consequently be seen as a deconstructive move. What it deconstructs is species supremacy but it also inflicts a blow on any lingering notion of human nature, *anthropos and bios*, as categorically distinct from the life of animals and non-humans or *zoe*. What comes to the fore instead is a nature-culture continuum in the very embodied structure of the extended self (...). (66)

Joy Harjo in *A Map to the Next World* reflects upon the turns of society during the last decades, on the evolution of the world and the revolution of technological advances. Although she is aware of the fact that power and economy have become the moving impulses of humankind, she hopes for a new horizon which awaits when population will realize its mistakes and step forward towards a more humane future in which justice and equity among the different species on Earth will finally prevail. The title suggests that instructions are given to the reader to find the exact geographical location and temporal dimension where humanity is heading; however, the truth is that Harjo aims more at bringing awareness towards the need to preserve our planet by criticizing and changing destructive attitudes that characterize human behaviour; thus, providing agency to each living being and opening the possibility of finding a better path that leads to create more compassionate worlds in the near future. She hopes for the Earth to recover its original established plan for all species that human ambition has been destabilizing for centuries and, in the case of America, to dismantle the new rules imposed by colonization in the past which have been serving to dominate and domesticate the landscape and its inhabitants, both human and non-human others. In an interview with Laura Coltelli, being asked about whether it is true Natives have a special sensitivity towards the environment, Harjo expresses her opinion about the engagement of Native American people with the conservation of the Earth "… generally women were, are, recognized as physically, electrically, whatever, more grounded, in tune with the earth, and again, that's a generalization, because there are always exceptions" (60). Avoiding essentialism, this statement also illustrates, as I attempt to make clear in the analysis of her poems, Harjo's cooperation with the natural world and her expression of a sense of place that she carries further to represent the map of her life, to develop her journey towards regeneration and her optimism about a radical change into a more cooperative attitude of human beings in a close future.

2 "A Map to the Next World" or How to Reconstruct the Past for the Future

The narrative poem that gives title to the collection focuses on the exploration of the history of Native Americans, since as Robin Riley affirms "history and myth are sources of knowledge and commitment" (150), because in Harjo's writings, most likely as a consequence of her mixed blood origins, the physical and the spiritual realm become one and the same thing and "her vision of a world in motion, with its layered landscapes and permeable boundaries, its multiplicity is thoroughly heteroglossic and dialogic" (150). She does not accept the separation between this world and other possible worlds that have existed or will appear: for her, the notion of timelessness impregnates the history of her people. In this poem, Harjo encourages her people to draw a new map but not to start from scratch by destroying everything that has already been built; on the contrary, Native Americans must search into their past to find the traces of ancestry, to incorporate important landmarks to it, whether they are physical, mental, temporal or geographical, such as those of language, in dialogue with non-human others but avoiding anthropocentrism. They must incorporate to their search the understanding of the change produced in tribal life from open lands and reservations towards life in urban contexts without forgetting the memories of tribal celebrations in ancient ceremonial sites.

This journey of the memory she undertakes in the poem towards regeneration and recovery, as it has been mentioned above, is absolutely circular, it starts and ends at the same point and it definitely has a future that must still be delineated: "Crucial to finding the way is this: there is no beginning or end" (Harjo 21). Apart from describing natural landmarks, she focuses on the presence of human and non-human beings; thus, she uses parts of the body, significant parts of the anatomy of living beings, as metaphors to explain her understanding of the world as living matter in constant variation; "The map can be interpreted through the wall of the intestine—a spiral on the road of knowledge" (20). The intestine, besides resembling a road in shape, also serves organisms to digest food, like Harjo, metaphorically, would like her people to digest the events of the past and profit from some memories while rejecting those which could damage elements from the universal map she is carefully creating. Harjo creates her personal map which appears included in the poem as well; it is definitely a mirror image of her search for her roots through language; it is the map of her conflictive identity as a native since she grew up facing rejection as a mixed blood. In sum, it represents the journey of her life, her own process of reaching acceptance and transformation at the same time. This reflection situates Harjo's writings once again close to Ortiz who also considers "place" as more than a geographical, physical or even spiritual concept: "Place is the source of what you are in terms of your identity, the language you are born into and what you come to use" (Ortiz 105). Language, the anthropological tool par excellence, as Braidotti understands it, is used by Harjo in this collection to displace the human being as the most important species on Earth. For example, in "humans aren't the only makers of poetry," she explains how the banana tree communicates by making poetry, by singing a song and also by making translations that are not necessarily conceived as linguistic ones. Plants are not conferred human characteristics but described acting as they are supposed to; they do what they should: drink water to grow. Harjo concludes the poem with a prediction and a warning for human beings: "we aren't the only creatures or the most likely to succeed" (112). The poem, then, offers more clues as to how our species cannot overlook other vegetal or animal species since they also share our space on Earth and, as Harjo announces, we humans do not seem to be aware of our fragility as living beings. It is for this reason that Harjo states that it is very likely that they are better endowed for survival. "Earthly Desire" also deals with the topic of sense of place and the importance of the transformation of matter, of energy in the universe. Harjo, conscious that she may be describing the ocean and the sky as humans, explains her reasons for such identification by affirming that she is highlighting the qualities that all earthly elements partake: our condition as material, organic beings. Even the sun, the ocean, the sky all contain biological qualities to be taken into account in the composition of the world, thus, bringing the gap between science and arts.

3 Aldo Leopold and the Land Conceived as an Organism

Men on the Moon by Ortiz completes this vision of the human being as in a constant search for new physical and ideological lands while forgetting the protection and conservation of the Earth and the connection of this topic with the development of a personal identity; thus, he introduces in his short stories, together with traditional elements, the reflections of the elderly people of his tribe who seem to be the only ones who have comprehended in thought and scope the effects of the modernisation of their surroundings into more degraded spaces where other priorities have dramatically transformed the rural and urban landscapes alike. Because of his knowledge of Spanish language, his title may suggest as well how distant humanity seems from reality, how lost and distracted our humanity appears currently, how we turn our backs on the future of the Earth and would rather be looking for knowledge "on the moon," completely absent minded, than worry and work for the improvement of the land we inhabit. Joni Adamson introduces "the middle place" when she discusses the work of Native American writers and defines it as "that contested terrain where interrelated social and environmental problems originate," and claims it as a space where "to work for transformative change" (xvii). The Earth that Ortiz attempts to recover looks similar to the next world that Harjo envisions in her poems; in fact, it seems to be the representation of this middle place at work to deal effectively with the current environmental problems and find a solution to avoid the ones to come in a near future.

These are writers who follow Aldo Leopold's theories of conservation and prove that they are still necessary even though they were developed almost a century ago in *A Sand County Almanac*:

> conservation is a state of harmony between men and land. By land, it is meant all things on, over or in the earth. Harmony with land is like

> harmony with a friend; you cannot cherish his right hand and chop off his left. That is to say, you cannot love game and hate predators; you cannot conserve the waters and waste the ranges; you cannot build the forest and mine the farm. The land is one organism. Its parts, like our own parts, compete with each other and cooperate with each other. The competitions are as much a part of the inner workings as the cooperations. You can regulate them—cautiously—but not abolish them. (189–90)

Ortiz and Harjo also agree in their writings with Leopold's affirmation that the discovery of the complexity of the land as an organism can be considered the scientific event of the 20th century, for indigenous cultures had always defended this fact. This land ethic they display, however, appears to be based on an ecocentrism as Buell defines it: "the interlinkage of the organismal and the inanimate" (137), and in the words of Leopold "the land ethic simply enlarges the boundaries of the community to include soils, waters, plants and animals or collectively: the land" (239). Giovanna Di Chiro also expresses her views against the concerns of environmentalists who are "obsessed with preserving and protecting those 'wild and natural' areas defined as places where humans are not and should not be in large numbers" (300), and defends the interaction of all natural elements, humans and animals included, in preserved areas. As these stories and tales by Ortiz and Harjo prove, most Native Americans had already understood the connections of all animate and inanimate beings before that date, and Leopold's theories come to sustain it. These are exactly their teachings. In fact, both writers have already experienced a change of consciousness in their lives: a change that has been undergone as a consequence of watching their people suffer the loss of land, displacement and, consequently, a significant loss of identity, so they can write about how it happens, how it comes to be. The Earth, the Sun, the Orb, all elements of nature and the universe partake in that transformation of conscience. Very much in consonance with what I have been arguing in the previous analysis of Harjo's "A Map to the Next World," Donelle Dreese discusses Ortiz's concept of land and the earth as united to identity and to human life, "and that is the sense of place which defines a person's life. Understanding the bond between identity and the land is crucial to understanding nature's significance for Native cultures" (93). As well as Ortiz, Harjo also portrays an image of hope for the human race embedded in the landscape, in agreement with the conjunction, of the forces of the universe.

> Maybe the spirits of the days live in the world above the earth we call dawn and stand reverently as the sun arrives in the appointed place in

> the east. Flowers, plants, all creatures prepare for the sun's arrival. You can hear the rustle of anticipation as the ultramarine sky turns to a cobalt grey. Deer move towards water to drink. Birds chatter plans for the day. In the distance a few humans begin running towards the east to give the sun encouragement, to make sure the sun returns to bless them. (59)

Harjo talks about leaving the body and the spirit flying to distant lands and spaces but she also makes a connection between matter, energy and human bodies, how they all contribute to the chain of life. This is clearly exemplified by Harjo in the story "twins meet up with monsters in the glittering city," when she talks about the genetic inheritance one of the characters has received: "death however does not end a journey and Greg will always look for his father even though his father lives in his muscles, his smile, his urge to live by water" (51). Harjo and Ortiz include in their writings personal life experiences that have contributed to the making of a significant change in their lifestyles, as Joni Adamson states when she discusses Ortiz's work:

> The Acoma oral tradition encourages the people to remember and learn from the past. Their uses of this tradition illustrate that human cultures, though blamed by some environmentalists and nature writers for the 'death of nature,' can play a role in teaching humans to observe and remember, to share their experiences, and, most importantly, to restrain themselves. (57)

In the story "when my son was born," Harjo takes time to review her teenage years when she was pregnant and remembers her thoughts back then in relationship with her change of conscience. As she is carrying a baby, she wonders why human beings keep destructive attitudes towards the environment: "As I walked through that Cherokee town I wondered what would happen to us, where humans fit on the evolutionary scale. Were we truly necessary to the survival of the biosphere? For a biosphere like earth to thrive each life form must reciprocate the gift of life. What do humans add besides stacks of trash and thoughtlessness?" (108). It seems a pessimistic thought but, on the contrary, it announces the necessary change that human beings must undergo, while at the same time it provides the tools for urgent improvement. All the stories together work as a manual telling us how to begin a new era; "the problem then is how to bring about a striving for harmony with land among a people many of whom have forgotten there is any such thing as land and among whom education and culture have become almost synonymous with landlessness. This is the problem of conservation education" (Harjo 210).

Harjo also addresses the issue of post anthropocentrism but this time in relationship with the animal world, in "the crow and the snake," a story which reflects on the equality, interrelationship and interdependence of species. The arrogance shown by the human family protagonist of the story in the interaction with the true protagonists: the crow, the dog and the other birds, sadly sheds light to the fact that the human species has forgotten how to deal with other living beings and has been historically imposing their laws as the unique norms to be followed by all creatures on Earth, rejecting any type of dialogic relationship without an anthropological basis. Harjo focuses on the lack of empathy presented by the human species and thus, tells the story from several different points of view, providing agency, but not anthropomorphisation, to the other animal species and highlighting the shrewdness of the crow. All the animals understand each other and plot to eliminate the snake, the alien element that appeared in the backyard of the house. Again, Harjo reflects on language when she establishes a comparison between human and bird language, in the similarities she finds between the way these animals communicate, especially the crow "whose language was closer to human languages" (32). Following Plumwood's dialogical interspecies ethics, human beings always attempt to establish communication with the more than human world in their own terms. The story projects the need that humans experience to subdue the non-human others and Harjo expresses this hierarchy established by the human protagonists with a deep irony. As the humans own the property, they feel entitled to throw away any other animals who would also like to enjoy the landscape and who don't understand what property rights are. The birds succeed in eliminating the snake when the crow risks his life to confront it and discovers that humans have tried to trick them with a plastic snake. The ending seems hopeful as every piece by Harjo; in fact in the ending of this story she depicts a posthuman scenario where new creatures finally interact and play together as well as listen to the story of the snake and the crow. She already foretells an ideal future in harmony for the young generations.

4 "Man on the Moon" or Stories about the Interaction of Nature and Technology

Man on the Moon deals with the impact of technological advances in Native American society. Ortiz reflects on the arrival of the TV to the reservation and the first step of humanity on the moon. Both events seem to coincide in time as if to show the idea of the human search for new challenges and new frontiers:

the human thirst to colonize new lands. Ortiz questions the idea of progress embedded in these technological advances with the perspective of time and contemporary readers can also deduct how the change produced after the introduction of space travels and the TV has not implied an improvement in the future of humanity or the Earth. When machines seem to have taken over the world, as in the example of the TV arriving to the house of the old man, then, natives prove that interaction with the surrounding environment is still the answer, the door open to produce a necessary change in mentality which comes before the transformation of institutions takes effect.

Irony and parody accompany the switching on and off the TV in the stories by Ortiz and a deep critique to a system based on wage economy and which gives higher priority to the investment on the development of machines and technology than to assure the survival and welfare of human and non-human species. The grandfather reflects on human stupidity, on the man who kills life on earth and travels to the Moon to find out whether life is possible there instead of trying to improve the conditions of the natural element on Earth. The need to see the tiniest bit of life in space in order to believe in the possibilities of a future for humankind makes the old man laugh; he is a sort of visionary who understands that the destiny of humankind is to return to the old ways, to traditions, and abandon the conquering spirit that has taken men historically to explore and take possession of new lands and places for the sake of submitting and owning them and their inhabitants. Only by taking a step back in the evolution of humanity, will societies show an attitude of respect as Leopold announced: "In short, a land ethic changes the role of Homo Sapiens from conqueror to the land-community to plain member and citizen of it. It implies respect for his fellow-members, and also respect for the community as such" (240).

As if the moral of the story was not clear enough, Ortiz retells the same event in "What Indians Do." This story also deals with the fact of human beings losing perspective and forgetting the point of view of magic, the magic of the natural way; how the human tendency has become a passive one; to develop a view as an outsider, as contemplating things happening around the world seated in front of the TV, instead of making them happen. Thus, he teaches the attitude human beings should adopt towards the world, the contrary to a passive one: a collaborative and cooperative one to make new things happen: "Because of my Western American education, I was looking into the square world of that box from so far away. I was watching, not participating in the event of my father speaking to me" (Ortiz 131). This is oftentimes our passive response to the degradation of life on Earth, to the lack of opportunity to give our opinion in places where it could mean a change. We contemplate the environment as mere audience, as recipients of an already made product. Dealing with attending as an audience to an event of storytelling, Ortiz returns to the characters of the story "Men on the Moon" where the grandfather is watching the arrival of the first men to the moon on TV but from a different perspective: this time he is an outsider, he is the storyteller within a different story. He describes again how the spacemen were gathering rocks to study them, justifying the trip as a search for knowledge. The grandfather, conscious of the enormous importance his grandchildren are conferring to the event, picks a rock from outside and shows it to his grandchildren stating: "That's knowledge!" (Ortiz 132). The children feel tricked but, in fact, he is teaching an enormous lesson to the younger generations: he brings them back to Earth, he reminds them that we have forgotten to care and communicate with the environment, with our surrounding nature which we don't stop destroying but we take pride on the fact of exploring and conquering other lands instead. As the story continues, Ortiz makes another reflection about a piece of news that appears in the newspaper where he finds the story of Uncle Sam that the Board of Realtors is sponsoring and distributing in schools during property week. Uncle Sam and the Board of Realtors embody the concept of private property that contrasts with the poet's claim for open spaces to be shared by nature and humans, by all living beings equally. Telling the truth about Uncle Sam has a double bind, a double meaning for an Indian who complains about the transformation of Indian lands and water into big corporations. Walking along San Francisco he runs into a lumberyard and the touch of the material transports him to his homeland, to the memory of a time where he could enjoy the experience of running free:

> I don't think anybody noticed the forests in me, the quiet footsteps I have taken in the Rockies and the Smokies. I don't think anybody knew the memory of touch in my hands of the trunks of great firs and pines and spruces… Longing so hard for forests and clean and gentle mountain wind, I snuck up 18th Street, a memory of white pine needles cradled in my arms like a baby. (136–37)

Ortiz understands that change must start within oneself, with a change of attitude towards the land, towards other living beings and the environment at large. He communicates with an older Acoma woman who mentions the concept of home as related to the Acoma Pueblo even though she has been living somewhere else for thirty years. Thus, the story ends with his own performance at The Indian Culture Day when he reads poetry, tells stories and sings while the younger generations play and laugh enjoying themselves. Ortiz, then, starts to envision a better future for them and his song becomes positive, optimistic, while he repeats: "I know that life is/I know that life is good" (139). Words and singing have brought all of them back home and the older generations mix with the newer and "they continue. They continue" (139).

The symbolism of the story "To Change Life in a Good Way" has been extensively analysed by Adamson who addresses it from the point of view of the positive transformation of Ida's garden, which goes from a barren start to a fertile production when she understands that "humans must recognize a reciprocal relationship with the land" (109). At the beginning of the story, Ida experiences a deep frustration because she is unable to grow anything in her garden, so, the locals, Mary and Pete, teach her how to deal with the soil. Finally, with Mary's help, she ends up breaking the clay and obtaining a good crop. Fertility goes hand in hand with learning to communicate and understand the land you inhabit, with a process of making it your home. Going a step further in the implications of this story, it is necessary to mention the importance of Slirk, Ida's brother in-law and a soldier in the American army who is killed at the front after stepping on an American mine in Vietnam. Contemplating the irony of his brother's death makes Bill doubt about the efficacy of governmental policies to improve the world and the life of American citizens and, by extension, about the basis of the American nation. Bill suffers questioning what his brother died for until he understands that it happened because he was too vulnerable like himself, without the power to decide his own future. Ortiz explains to his readers that the hierarchy established favouring progress in economic terms while destroying nature does not benefit humans or other living beings and an alternative lifestyle must be searched for. He introduces this idea of regeneration through Pete's offering:

> He said, This is just corn, Bill, Indian corn. The people call it Kasheshi. Just a dried ear of corn. You can take it with you to Oklahoma or you can keep it here. You can plant it. It's to know that life will keep on, your life will keep on. Just like Slirk will be planted again. He'll be like that, like seed planted, like corn seed, the Indian corn. But you and Ida, your life will grow on. (113)

At the end, when Ida obtains the first produce from the garden, the positive thought that Slirk didn't die in vain appears. The whole family discovers the importance of learning to appreciate and protect your surroundings. Living in a mobile home, they have become used to being constantly on the move ignoring the effect of the human footprint on the landscape. They finally realize that respect for the environment and care for the Earth means a difference in attitude that confronts the sadness and violence of war, the ways of capitalism which bring destruction instead of creating knowledge as nature does.

In the story of "Feathers," in which a father and a son adopt a stray cat, Ortiz establishes a comparison between the lives of the man and that of the cat; both species, the human and the feline, seem connected and equated. They receive the same treatment from the child and his mother as they appear to have been previously abandoned and left in the street. Thus, the cat functions as the metaphor of rejection or acceptance: when the father expresses his desire for the mother to keep the cat, in fact, he is wondering whether he will be thrown away as he has been experiencing for the last years. But the woman invites both in and apparently lets them stay. It could be thought that the cat makes the miracle of uniting the couple again; that the parents get reconciled thanks to the new life that is brought to their household and must be taken care of. However, it is the whole family who learns to see things in a different light at the arrival of Feathers. The mother shows respect for the poems the father has published and for the cat, and keeps it instead of giving it away as she had done previously with a puppy, understanding that harmony and solidarity benefit everyone. As well as Harjo, Ortiz also teaches respect for animals and how to value other species as different and not inferior, removing hierarchical structures and the standard of man as measure of all things as Plumwood and Braidotti defend in their ecofeminist theories. The death of the dog named Old Horse in "The End of Old Horse" brings awareness about respecting life and all living beings. The children laugh at the dog tied up making efforts to escape because they don't think he is in trouble or appreciate he is suffering. When they are told the news that he has choked to death trying to run away, they realize the cruelty of both actions: tying him up and laughing at him. A similar teaching comes from the death of George, the goat that refuses to obey human laws and is tied to a post in the story titled "Distance;" George struggles to free himself from the rope and the post unsuccessfully. Then, he spends several days and nights outside in the heat without being given food or water. The suffering of the goat is clearly stated: "On shaky legs, George bleated, and the sound was pleading" (165). Ortiz denounces through these two stories the abuse and material exploitation of animals when they are treated as possessions, clearly objectified instead of being respected as living beings. It is children in these stories who show sympathy and teach adults to be compassionate with other species. Adults attempt to tame children and animals equally through punishment but the little ones learn to value non-human others and express their understanding of their suffering. Their emotions when the animals finally die contain their rebellion against animal abuse. At the beginning of the stories, they trust adults, they rely on their teachings but, little by little, as they develop a bigger empathy with the animals, they lose innocence and criticize the decisions of the grown-ups. Unlike adults, these children have already experienced a change of conscience towards a new land ethics that eliminates hierarchies and includes other than humans in its configuration.

5 Conclusion

Collaborative and creative work among human beings, the necessity to cooperate labouring together to make transformative change effective summarizes Harjo and Ortiz's aims in the literary texts analysed here. They do not give up revealing the problems the world faces but they adopt a very affirmative attitude towards them, being confident about the possibilities for improvement that they offer in their literature and providing numerous and practical solutions.

Drafted from personal or communal experiences, the events narrated and the humans and non-human others portrayed always show an awareness towards the problems of their surroundings that teaches the younger generations about the importance of eliminating hierarchies of any kind, rejecting human supremacy among vegetal and animal species. Thus, it is through children or non-human others that the messages are transmitted. They also sustain the need to transform anthropocentric attitudes into more dialogic interspecies ethics that include solidarity and recognition of the interdependence of species on Earth.

Harjo and Ortiz display a more than evident post anthropocentric alliance with the more-than-human world. Their work is hopeful and optimistic as they envision a positive change in human minds and a new philosophy based on ecocentrism that will certainly improve the conditions in which life is developed on Earth. The authors' imagination creates new maps which inspire readers, illustrating a positive relationship between the human and the more-than-human element at the same time that they provide an explanation of the Native American cosmovision that encourages dialogical encounters and sustainable developments.

Works Cited

Adamson, Joni. *American Indian Literature, Environmental Justice and Ecocriticism: The Middle Place*. U of Arizona P, 2002.

Braidotti, Rosi. *The Posthuman*. Cambridge: Polity, 2013.

Buell, Lawrence. *The Future of Environmental Criticism: Environmental Crisis and Literary Imagination*. Blackwell, 2006.

Coltelli, Laura. *Winged Worlds: American Indian Writers Speak*. U of Nebraska P, 1990.

Di Chiro, Giovanna. "Nature as Community: The Convergence of Environmental and Social Justice." *Uncommon Ground: Rethinking the Human Place in Nature*, edited by William Cronon, Norton, 1995.

Dreese, Donelle N. *Ecocriticism: Creating Self and Place in Environmental and American Indian Literatures*. Peter Lang, 2002.

Harjo, Joy. *A Map to the Next World: Poems and Tales*. Norton, 2000.

Harjo, Joy. *How We Became Human. New and Selected Poems 1975–2002*. Norton, 2004. Kindle edition.

Joy Harjo. joyharjo.com. Accessed 26 Feb. 2021.

Leopold, Aldo. *A Sand County Almanac: with Essays on Conservation from Round River*. Ballantine, 1970

Martín Junquera, Imelda. "The Inscription of the American Southwest in Navajo Tribal Parks." *Iperstoria: Testi Letterature Linguaggi*, no. 9, 2017.

Ortiz, Simon J. *Man on the Moon: Collected Short Stories*. U of Arizona P, 1999, pp. 90–8.

Plumwood, Val. *Environmental Culture: The Ecological Crisis of Reason*. Routledge, 2002.

Riley Fast, Robin. *The Heart as a Drum: Continuance and Resistance in Native American Poetry*. U of Michigan P, 1999.

Westling, Louise. "Literature, the Environment and the Question of the Posthuman." *Nature in Literature and Cultural Studies: Transatlantic Conversations on Ecocriticism*, edited by Catrin Gersdorf and Sylvia Mayer, Rodopi, 2006.

CHAPTER 8

When Matter Takes a Position

Post-anthropocentric Landscapes in Contemporary Art

Bárbara Fluxá Álvarez-Miranda

Abstract

In response to the systemic global crisis a new cultural paradigm is taking place as an imaginative alternative the hegemonic modern narrative, displacing humanity from its persistent anthropocentrism, under which the instrumentalization of the world and its materials has been justified exclusively in materialist terms. This network of thought advocates for a new subjectivity fueled by the hybridization between nature, techno-science and culture, where contemporary art acts as a source of critical knowledge to establish a transdisciplinary approach towards the understanding of this paradigmatic setting. Many artists like Olafur Eliasson, Ursula Biemann or Regina de Miguel for example, in collaboration with specialists in climatology, hydrology, oceanography, and especially those who work in the study of the solid Earth—geology, geography, soil sciences or geodesy—go beyond the simple representation of nature to propose multidisciplinary complex landscapes grounded in post-anthropocentric affect and emotion.

Keywords

Matter, post-anthropocentric, contemporary art, landscapes.

1 Contemporary Paradoxes

Humanity is experiencing a paradoxical historic moment in which it is difficult for us to accept a positive view of ourselves as a species because while planet Earth is suffering the greatest systemic environmental crisis in our history (IPCC),[1] we continue to make enormous progress in the fields of science,

1 Climate change has been ratified by the IPCC. The Intergovernmental Panel on Climate Change (IPCC) was created in 1988 to provide comprehensive assessments on the state of

© KONINKLIJKE BRILL NV, LEIDEN, 2022 | DOI:10.1163/9789004501270_010

medicine, and technological innovation (Harari 31).[2] On the one hand, we develop incredible technical and scientific means that are capable of resolving a large part of the planet's ecological problems and social imbalances, and yet, on the other, we are not capable as a subjective and affective social force to put them into practice (Guattari 42). This *modus operandi* occurs because the Western tradition has been based on mistaken ethnocentric beliefs about the superiority of human beings over other beings, the universe, and its matter. Paradoxically, although the nature/culture dualism lacks an ontological basis, it has developed over the history of humanity until becoming reality, with the ironic result of it finally dissolving as a result of today's complete dominance of nature by man.

From an ecosystemic point of view, Earth science specialists tell us that over 90% of plant growth currently takes place in agrotechnology systems run by humans; 90% of all biomass – of all living things – is originated by man or domestic animals; and over three quarters of the Earth's surface (free from ice) is no longer in its "original" state, but rather in cultural landscapes created directly or indirectly by man (McNeill 260). Based on an infinite number of intergovernmental studies and reports, today it can be objectively asserted that human activity with its social and economic processes has become a significant factor for change and transformation in how the Earth functions, on a global, systemic scale, on par with a geological or climatological agent. Furthermore, the existence of a new atmospheric layer on our planet called the "technosphere"[3] can be confirmed: a new anthropogenic planetary sphere which refers to the technological-functional palimpsest complex into which we have converted our habitat's ecosystem. This is the premise of the scientific theory of the Anthropocene,[4] which affirms that the Holocene, which until now had been our geological era, has ended. However, rather than enter into

scientific, technical, and socio-economic knowledge on climate change, its causes, possible repercussions, and response strategies. See: www.ipcc.ch/home_languages_main_spanish.shtml (accessed June 2018).

2 To mention just some achievements: in medicine, for example, of note are the increase in life expectancy and control over disease (except cancer and HIV); in astronomy, greater knowledge of the universe thanks to the launch of satellites and rockets to outer space; or in technological innovation, the creation of electronic devices that have brought our lives into an intangible world that is hyper-connected thanks to information systems and the digital communication of a networked society (Harari).

3 The neologism "technosphere" (now used fairly frequently) was proposed in a somewhat premonitory fashion, if I may say, by the geochemist Vladimir Vernadsky (1863–1945) in his famous book *The Biosphere* (1926).

4 Term coined by Paul J. Crutzen, Nobel Prize in Chemistry in 1995 for his studies on the hole in the ozone layer.

scientific debate, what I would like to remark here regarding this incredible theory is that its mere existence – as a multi-disciplinary world view of the human race as preying on its own habitat, until fully destroying it – is resulting in society's profound awareness and an ethical revision of the Western way of inhabiting the world, especially because the driver of Anthropocentrism and instrumental modern scientific thought is based solely on the myth of progress. The good news is that, precisely thanks to this unpleasant awakening, we are finally beginning to realize to what extent we are inseparable from the system and how our dominating action is affecting its destabilization.

That promising economic, scientific-technical, and political undertaking of Modernity – universalist, ethnocentric, imperialist, and colonialist – has failed; as has the anthropocentric thought that gave free reign to planetary domination based on a relationship with natural "capital" exclusively in terms of economic exploitation (Mignolo 48). For too long, we have managed our world through the needs and urgencies of the capitalist market, which have materialized in the modern world as a consumerist delirium of production and destruction of space and nature. The management of what today is referred to as "second nature" (Morton 82) drifts between conflicts of outrageous (unplanned) interests, at the whim of political power.

2 A Transdisciplinary Network of Post-Anthropocentric Sensitivities

Given this context, the new humanities, bridging the traditional gap between science and the arts, defend the disappearance of the historic longing of the modern subject, who, located at the center of the universe, dreamed of a world stripped of surprise, which would become a mere technical automatism. The new human and social sciences are shifting away from the "man" of Modernity and, before that, of Christianity and the Renaissance. It seems that in our time, "it is not becoming the planet's stewards that Earth-lovers dream, but of a time when humans ceased to matter" (Gray 17). Contemporary philosophy speaks of a man who is born with scarcity and indeterminacy (Castoriadis 193), which forces him to culturally channel the empirical and sensory experience, to regulate it through knowledge, intuition, and his senses, producing the world. This individual man must re-balance relationships with his own subjectivity (mental ecology), with society itself (social ecology), and with nature (environmental ecology) to establish a new way of producing the world (Guattari 56).

Today, a new cultural paradigm is being forged as an alternative to the modern hegemonic model, which aims to shift humanity away from its anthropocentrism, and advocates for a new subjectivity full of hybridization and interferences between nature and culture (Latour, *Nunca* 55). Its challenge now is to find a new formula outside of the historical hegemony of humanism, in order to offer different paths and hypotheses to put into practice a new human behavior that is interconnected with the planet (Braidotti 39). It seems reasonable to think that this will require of us, at first, an enormous effort to reconnect with what is most essential, our habitat, in an attempt to "reset" (GLOBALE)[5] our relationship with nature. To do so, we must define new concepts that signify us in all of our complex reality, through a process of understanding that overcomes the culture-nature dualism and which, in turn, will allow us to reformulate solutions to our current problems. We must overcome the separation between the boisterous subject of law–represented by politicians in parliament–(Serres 143) and the silent object of science–represented by scientists in the lab–in order to produce a new space where new mutual understandings can be established between the power of social force and the mechanisms of natural force, as Bruno Latour argues:

> If [on the contrary] our Constitution authorizes anything, it is surely the accelerated socialization of nonhumans, because it never allows them to appear as elements of "real society." By rendering mixtures unthinkable, by emptying, sweeping, cleaning and purifying the arena that is opened in the central space defined by their three sources of power [nature, society, absence of divinity],[6] the moderns allowed the practice of mediation to combine all possible monsters without letting them have any effect on the social fabric, or even any contact with it. (Latour *We Have Never* 42)

It is precisely this lack of contact to which Latour refers that the multiple initiatives of critical thought and reflection aim to correct. Hundreds of independent individuals and intergovernmental community organizations of all types have added their studies to scientific contributions in order to join forces and generate a space for interrelations given the urgency of the situation. This reintegrating paradigm shift is being undertaken in collaboration with the humanities,

---

5 All of these issues are being tackled by several contemporary artists and curators in their projects. Suffice it for now to mention the exposition entitled *GLOBALE: Reset Modernity* curated by Bruno Latour in ZKM | Center for Art and Media Karlsruhe in 2016, from where I extract the idea of using the word "reset" to allude to the fact that nature needs to be restarted as if it were a computer system in order to return to its starting point. See: www.zkm.de/en/event/2016/04/globale-reset-modernity (accessed June 2018).

6 Clarification from the author to specify, in the context of the quote, what three sources of power Bruno Latour is referring to: transcendent nature, a free society, and the absence of divinity established by Modernity thanks to the Constitution.

aesthetics and the arts, the sciences, and economics so that our relationship with nature may be established not on antagonistic terms, but rather complementary terms (Prigogine and Stengers 293). This is the case, among other reasons, because these initiatives hold the fierce conviction that a new system of ethical values must be proposed in order to definitively eradicate the old humanist ideas that have led us to the Anthropocene and, at the same time, the economic and political strategies of the Plantationocene and Capitalocene:[7] colonialism, imperialism, technocracy, globalization, and Eurocentrism (Haraway 16), which allow the world to be conceived in exclusively materialistic terms and based on the center-periphery dualism. Only through this ethical change in society's conscience in the face of global crisis, they say, can we rebuild our behavior towards our habitat and thereby improve the quality of life of our own species. And by improving the lives of human beings, we will improve the lives of the living beings, artifacts, and matter that surround us, the *nonhumans* (Latour, *Políticas* 332) guaranteeing a balanced future for the development of planet Earth and the laws and energies that govern it. So then, although somewhat late, after several decades of activism, it seems that this new ethical dimension of respect for the environment is here to stay, as are the principle of responsibility towards future generations, concern about preserving the planet, and international and intergovernmental consensus with regard to the need to take urgent action in the face of the new reality of climate change and the impact of our social-economic model on the planet.

Now then, how to achieve this balance? From what conception of the world should we approach our existence? How can we confront the conflict between the finiteness of natural resources and our contradictory technical system based on unceasing, irresponsible consumption? It seems that there is a certain consensus within these new currents of thought, in which the lines of action and governance for society, looking towards the future, should aim not to preserve or recover a given form of nature–which no longer exists as such–but rather to achieve a new planetary balance among all the spheres that

7 The terms Plantationocene and Capitalocene were popularized as part of the new cultural studies by Donna Haraway in her celebrated text *Anthropocene, Capitalocene, Plantationocene, Chthulucene: Making Kin* (Haraway, 2016); however, they had been suggested earlier by the anthropologist Malm Moore. In any event, the two terms are used to underline the fact that the (ethical and political) responsibility for the ecological and climate crisis of the Anthropocene is not shared by all human beings equally. Not all humans have had the same social and political power to influence the capitalist economic system that underlies said crisis: not the slaves of the development of European capitalism in Southern America, nor the nearly-exterminated and marginalized indigenous communities, nor the people who suffer the misery and labor exploitation of the neoliberal system (Moore, 2016).

comprise the world today: the atmosphere, geosphere, hydrosphere, biosphere, as well as the recently denominated noosphere and technosphere. To do so, a transdisciplinary global culture–that has been ethically agreed to (Guattari 35)–must be developed where the focus is on an urgent reconnection with this new natural-artificial system in order to provide coordinated global solutions that are the result of an understanding of the system in all of its complexity.

Thus, a new transdisciplinary culture is on the horizon, one that is committed to the survival of the second nature and man (Morton 49), and which prioritizes its ethical, social, and political principles above economic and materialistic values. This new cultural paradigm has space for a new transdisciplinary science that experiments and develops in the direction of technological innovation, but towards the promotion of a new model of knowledge based on the search for social and ethical values (Echeverría 160), because science without a social conscience or environmental ethics–solely at the service of the market economy and geopolitics–is a far cry from the discipline that the planet needs today.

The new cultural ethics that this collective social conscience should promote must be far from the notion of the instrumentalized nature of mechanical science and the capitalist economy. We need a new culture that finally accepts that human beings are interdependent on other living things, inert matter, and natural ecosystems and that, while cultural by nature, they are in no way all-powerful. Instead, humans must know how to negotiate with a space that, as contemporary science tells us, is indeterminate, perilous, in chaos, and based on the principle of uncertainty. Humanity must shift from its sole aspiration being to acquire "knowledge" of the world as a means of manipulating it through science to "understanding" the world through man's interaction, not as a dominating force, but rather as a component. The new science tells us that the universalist dreams of modern science have been toppled by phenomena such as magnetism and thermodynamics. As a result, we can no longer be absolutely sure of our knowledge, we cannot project ourselves into the future (except in a probabilistic manner), because reality can no longer be understood independently of time. Time in contemporaneity, understood as an essential factor in the dynamism of what is real, chaos, action, and chance, results in a renewal of the conception of the mechanist world, as the system is now seen through the lens of the unstable and dynamic. And, as difficult as it may be for us to accept it, this all implies a view of nature as unavailable and, in large part, untamable by man (Kaku 335). The humanistic belief that humanity is capable of controlling its destiny is increasingly far-fetched.

3 The Participation of Artistic Practices

The complex new spatial-temporal realities described above are also reflected in the narratives of contemporary artistic practices. Artists from around the world, immersed in these new transdisciplinary currents of thought, and drawing from their artistic imagination, are proposing a decentralization of our outlook on nature through new "post-anthropocentric" landscapes as an alternative to the performance of the traditional landscapes that were characteristic of humanism. They understand that there must be a new interrelation of coexistence with nature which destroys the classic subject/object dualism, which granted a central, advantageous position to human beings, compared to a fixed, passive nature at the disposition of our perpetual domination. They are putting into practice strategies of critical, ethical, and political art that are committed to modern society. They are revising the function of art and artists, encouraging a profound reflection on art's capacity to generate knowledge. They are making cultural mediation possible in society because they want to participate as active agents in the reconstruction of our world. They are inserting themselves into society to activate it, proposing their work as a short-circuit or wake-up call that will generate awareness, in most cases in the format of a multidisciplinary artistic project that goes beyond the work of art as an object exclusively destined for the art market.

Today's artistic practices interested in the issues of contemporary culture and nature establish transdisciplinary relationships with the new humanities, social sciences, anthropology, psychology, and phenomenology; but also, with the hard sciences related to this new world view instigated by both climate change and the Anthropocene: physics, chemistry, life sciences, and, especially, Earth sciences. The artists acquire and put into practice methodologies typical of these disciplines that are in line with their way of understanding the world as a strategy for artistic production. In this conceptual role swapping, in which the artist carries out a kind of division or shift towards the place of the "other"–that of the amateur as anthropologist or geologist, for example–the acquisition of methodologies characteristic of the scientific disciplines as a means of creation enters into play. Through scientific and technological means, they create visually and intellectually hybrid works of art that explore the material and cultural nature of our world, at the same time questioning how these devices mediate our experiences. Field work, producing taxonomies, compiling files, displaying experimentation devices, teamwork and collaboration, and the use of technical innovation and new technologies are the new strategies for creating their works. Through their work, they do not "depict" an image of a landscape, but rather they "present" material and immaterial elements–not necessarily of an aesthetic nature–from the world that surrounds us as an artistic mechanism. Objects, tools, artifacts, matter, documents, phenomena, sensations, ideas, and experiences are now the materials of the new "post-anthropocentric landscapes."

4 When Matter Takes a Position[8]

> One's mind and the earth are in a constant state of erosion, mental rivers wear away abstract banks, brain waves undermine cliffs of thought, ideas decompose into stones of unknowing, and conceptual crystallizations break apart into deposits of gritty reason. Vast moving faculties occur in this geological miasma, and they move in the most physical way. This movement seems motionless, yet it crushes the landscape of logic under glacial reveries.
>
> (SMITHSON 45)

Of all the transdisciplinary strategies of contemporary art proposed above, I will focus on collaborations between contemporary art and Earth sciences. Hydrology, oceanography and the disciplines involving the study of solid Earth (geology, geography, soil science, and geodesy) become potential tools for relativizing the insignificant place occupied by humanity, not only in the history of the geological time of planet Earth, but also in that of the entire universe. By interrelating with these sciences, and the presence of their matter, the artists propose a new system of ethical values with regard to nature that makes plain that not only are the old anthropocentric ideas that justified the exploitation and instrumentalization of the world and its matter exclusively in materialistic terms out of date; but they are also not valid because they spatially and temporally situated human beings in a place–at the center of the universe–that today we certainly know does not correspond to them. In response, now inert matter occupies an ironic central role in the representation of nature and the universe, thanks to a voluntary cosmic decentralization of the contemporary *anthropos*.

---

8 A play on words based on the essay by philosopher Georges Didi-Huberman *When Images Take a Position* on the poetic work of the playwright Bertolt Brecht. I have replaced "images" with "matter" because I find similarities between the poet's creative method and certain contemporary artists, in terms of the strategic use of dialectic montage as a language for representing the world: decentralization of the spectator, superposition of images/matter, complex transdisciplinary mediation, and multiplication of spatial-temporal perspectives.

I will now examine just a few examples of the multiple contemporary artists who are following these strategies for presenting the real and the natural. I will address some relevant works that are based on the appropriation of matter as the main focus of the contemporary landscape narrative. I will interpret this strategy as a committed, conscious proposal to displace the creator subject from the act of artistic performance so as to give way to the object–already provided by nature–as an eco-historic witness of the spatial-temporal reality of our world and, consequently, of the universe. As if by presenting the object/matter that makes up the real world as a non-anthropocentric, natural, ready-made element, it could represent itself without the need for mediation by the colonizer, dominant human being. Although this strategy's objective of displacing the inevitably subjective action of the creator subject may seem utopian, it seems to be legitimate and promising because of the goal it pursues. Utopian stories are still possible, just as "ideas decompose into stones of unknowing" (Smithson 124). Let us examine some of them, then.

5 *Arctic Sea Ice in* Ice Watch *by Olafur Eliasson, 2014–16.*[9]

This work, created by artist Olafur Eliasson (Copenhagen, 1967) in collaboration with the geologist Minik Rosing, directly addresses the issues of atmosphere and air quality as an ethical-political, as well as artistic concern. To carry out this project, Eliasson collects 12 large 100-ton blocks of ice that have broken off from the Greenland ice sheet from a fjord outside Nuuk and presents them in a clock formation in a public place open to citizens, while they creak in their melting process. The installation aims to raise awareness of climate change by providing a direct and tangible experience of melting arctic ice. It reminds us that not too long ago, perhaps a bit over a century, the prediction of time was directly related to the survival of the species, to physical wellbeing; and that, due to the technological inventions of the modern era, our relationship with the climate had become secondary, until the realization of climate change made us aware of this modern illusion. The unpredictability of time is a phenomenon that refuses to be fully colonized, and even more so now that global warming bears witness to this fact.

9 *Ice Watch* has been installed in two places as of now. The first installation was in Copenhagen, in City Hall Square, from October 26–29, 2014, on the occasion of the publication of the UN's IPCC Fifth Assessment Report on Climate Change. The second installation took place in Paris, in the Place du Panthéon, from December 3–13, 2015, for the United Nations Climate Change Conference, COP21. See: *www.icewatchparis.com/* (accessed June 2018).

6 *Glacial Ice in* Subatlantic *by Úrsula Biemann, 2015*

In this science fiction video essay entitled *Subatlantic* (web), the film artist Ursula Biemann (Switzerland, 1955) juxtaposes science, geology, and climatology with the history of humanity. The film's story takes place during the Subatlantic period, the last climatic phase of the Holocene which began about 2,500 years ago, and which has witnessed the majority of humanity's historical events, such as the Roman Empire, the discovery of America, and both world wars, in addition to many extinctions of all kinds of species. The voice over that is heard throughout the film is allegedly a female scientist who is making instrumental observations about the changing spatial-temporal landscape surrounding the last glacier thaw. As the video advances, the images situate us in an increasingly submerged place of oceanic observation, where the viewer examines both the objects of the physical world–the atmosphere that is enveloping her–as well as the thoughts of humanity itself, which are reconfigured or modified under the changing conditions of the new climatic scenario. *Subatlantic* also alludes to the submerged space of the Atlantic Ocean, from Disco Bay in Greenland to a small Caribbean island, showing locations that are very far apart but that, in turn, are connected by ocean currents. Biemann addresses a subaquatic landscape with dynamics of connection that are invisible to the human eye, a hidden, uncolonized place.

7 *Copper in* Copper Country (*Bingham Canyon, Chuquicamata, El Morro*) *by Maarten Vanden Eynde, 2016*

The artist Maarten Vanden Eynde (Belgium, 1977) is currently investigating the influence of the transatlantic trade of materials that are historically and culturally essential to Western civilization (rubber, oil, ivory, copper, cotton, and uranium), but also the trade of human beings. He is interested in how this trade has affected the evolution of contemporary society, the creation of nation-states and power structures on an international level, and, lastly, our own planet on a global level. The project *Triangular Trade* (web) traces the origin of different materials and follows their evolutionary path as they are processed and transformed into commercial "treasures." The artist is interested in highlighting the little-known and often overlooked global historical influence of the Western world's trading practices in the Congo through the exploitation of its natural resources, especially cotton, copper, and uranium, which played an important role in the "triangular trade" between Africa, the southern United States, and central Europe. In his installations, through a decolonial strategy, he merges the recovery of the historical and political memory of matter, the contemporary drifts of globalized trade and its colonization with the material remains of technological evolution.

8 *Calcium Carbonate in* The Mineral Body *by Ilana Halperin, 2013*

This series of sculptures was created by Ilana Halperin (New York, 1973) in a small thermal city in the mountains of Auvergne, France. There, several generations ago, a family founded the *Fontaines Pétrifiantes* to create limestone sculptures made through the same natural process by which the carbonated water of volcanic caves forms spectacular stalactites. Inside these limestone caves, there are carbonate-rich waterfalls that the artisans use to produce all kinds of petrified objects. Normally, in a limestone cave it takes stalactites one hundred years to grow one centimeter. In the *Fontaines Pétrifiantes*, thanks to a system of steps and casting ladders in heights, one centimeter grows in one year. The pieces in *The Mineral Body* (web) installation took four months to form. The petrification of this series of organic-looking wooden shapes astonishes the spectator, who is faced with calcareous objects/subjects of a strange hybrid nature. Halperin approaches stone not as dead matter, but rather as a living geological process. The mineral is presented to us as a fluid, living substance full of autonomous creative capacity, where human intervention is not required.

9 *Limestone in* El capítulo de los áridos (*CaCO³*) *by Bárbara Fluxá, 2018*

This audiovisual installation by the author, Bárbara Fluxá (Madrid, 1974), was created in the environment of an open-pit limestone quarry, in a small rural town in the valleys of Asturias, northern Spain, where the artist has family ties. The work (web) reconstructs–through videos, virtual 3D models, drawings, and photographs–a mountain that disappeared as the result of mining operations after over 40 years of activity. This time period coincided with the life of the artist, who watched the mountain disappear gradually throughout her life, explosion after explosion. In collaboration with an archaeologist specializing in virtual reconstruction of land and by recording the mountain being blown up in the course of ongoing industrial activity with several cameras, the artist poetically reverts—in the context of art—the emptying and breaking perpetrated on this familiar landscape. New technologies allow us to reverse human action in a dramatic gesture of utopian recovery. This work is part of a broader

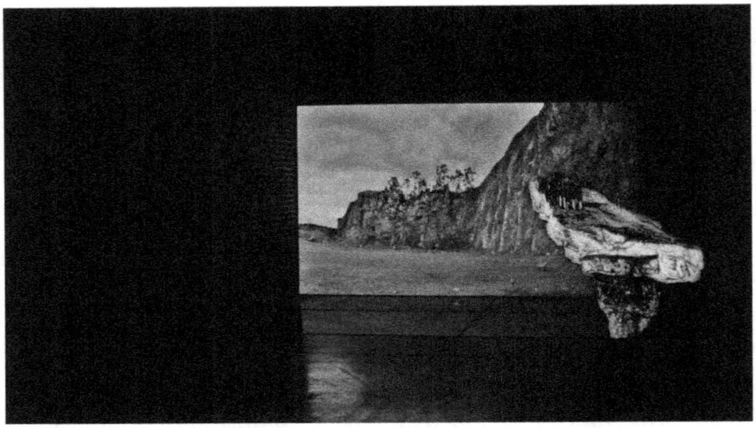

FIGURE 8.1 Still from the installation *El capítulo de los áridos* by Bárbara Fluxá, 2018.

project entitled *Capital Natural*, which reflects upon one of the great debates in modern society: the insatiable material exploitation of the land by the neo-capitalist economic system, whose only way of relating to the world is exploiting and dominating it. The work appeals to the debate regarding the necessary deceleration of hyper-consumerism and the unsustainable use of limited natural resources. It aims to point out the technological arrogance of contemporary human beings that enables them to make land that is millions of years old disappear from the geological map in a matter of seconds. The negative morphology of the contemporary Asturian landscape – from the bird's-eye-view of Google Earth – with hundreds of mountains that have been blown up is a good example of this problematic human domination.

10 *Granite in* Earthworks *by Semiconductor, 2016*

The video installation *Earthworks* (web), conceived by the duo Semiconductor (Ruth Jarman & Joe Gerhardt, England, 1972), at the La Planta quarry in northern Spain was created in collaboration with Spanish scientists. The work displays data on the glacial, seismic, volcanic, and artificial activity at that quarry, projected as a digital animation with sound on five screens. The translation of these data to sound and digital images represents, in some way, the activity of the Anthropocene, our new anthropogenic geological era. The images resulting from the seismic data are fluctuating, spectacular marbled waves full of color, layers that are animated with the sounds of earthquakes and volcanic, glacial, and human activity, registered as seismic waves. *Earthworks* offers an immersive experience to the spectator, who now sees–from inside the geological phenomena themselves, materialized in virtual models–the formation of the landscape through the same scientific and technological devices that are used to study it. Semiconductor is attempting to make visible both nature's hidden geological forces and human beings' tools and technologies for scientific measurement. These two forces possess two diachronic temporalities, the impossible-to-conceive geological temporalities which occur beyond the chronological time of human beings, and anthropogenic temporalities, which, in recent decades, have become visible and quantifiable with multiple scientific indicators. The artists defend the idea that the formation of the landscape can only be experienced through the technological mediation of nature. *Earthworks* draws attention to information about time, space, and the geological phenomena that transform the Earth, which allows us, for the first time, to be witnesses to the overlap of geological time with human time.

11 *Volcanic Rock in* Quatemary *by Tacita Dean, 2014*

Film-maker Tacita Dean, in her photographic polyptych *Quatemary* (web), shows us a post-apocalyptic geological landscape created using a blend of modern images captured by the artist and historical photographs from the 19th century of the emblematic Yellowstone National Park in the U.S. The work is a dystopian landscape fiction based on the hypothetical possibility of a volcano erupting and splitting the American continent in two, creating a dust cloud that would block out the sun and lead to the extinction of all living beings that inhabit the Earth. The resulting dystopian images produced using a strange aesthetic somewhere between the fantastic and the documentary depict a barren landscape as an enigmatic post-nuclear wasteland. This landscape attempts to show us everything that science believes was on our planet before the human race appeared, and what will most likely remain once we are gone. *Quatemary* is related to Dean's artistic concern with capturing memory, the spatial-temporal future, and coincidences and chance in the historical events that take place in human beings' lives.

12 *Seismic Quartz in* Medium Earth *by The Otolith Group, 2013*

In this film essay (web), the duo The Otolith Group (Kodwo Eshun, London, 1967 and Anjalika Sagar, London, 1968) explores the seismic activity of the geological landscape of California, U.S., as well as the psychic-sensory activity of certain "mediums" who say they can detect earthquake activity before it occurs. The main protagonist of the film is one of these "mediums" who is capable of predicting the electromagnetic field activity generated by tectonic plates during earthquakes through complex psycho-physical sensations. Through scientific and documentary images of the tectonic morphology of California, the artists explore the confrontation between the mythical cultures of geo-poetic prophecy and prediction that mediate the psychic and sensory experience of humans and the Earth and the hard rationalist sciences that limit themselves to viewing the physical-seismic shaking of quartz substrates translated in the data from their magnetometers. However, while the geological sciences do not accept these sensory capabilities in humans because they assert that our cerebral cortex has lost the ability to perceive and translate these signals, they do recognize that the limbic brain of animals can do so. Science knowns that earthquakes locally disturb the planet Earth's natural electromagnetic field, and that this micro "piezoelectric" disturbance of the quartz is detected and translated by the "natural magnetometer" of many animals' brains. Although it does not allow humanity to possess this sensory ability, it does endeavor to research technological systems that can predict these events for us.

13 *Dark Matter in* El conocimiento nunca viene solo *by Regina de Miguel, 2013*

The work *El conocimiento nunca viene solo* (web) by Regina de Miguel (Málaga, 1977) is related to this revision of the effects of the modern project on our contemporary society, as it offers an analysis of the different types of knowledge produced by science. Using a very special area of the Chilean geography, the Atacama Desert, as the subject, this visual essay constructs a narrative in which different types of knowledge intersect in one of the most extraordinary landscapes in the world. The extreme climatic conditions of this remote desert region make it almost impossible for life forms to live and proliferate there; but at the same time it is an ideal region for astronomical observation of the cosmos thanks to the limited light pollution and its geographic location on the planet. The central theme of this work is an examination of how scientific knowledge is attained in the West, as knowledge is a very powerful tool subject to specific geopolitical and socioeconomic agents. Compared to other areas of knowledge that involve implicit processes of self-criticism and which are aware of their political dimension in society, scientific disciplines have built their narratives with an independence which, in addition to safeguarding an illusion of autonomy with regard to the reality that supports and drives them, has allowed them to exclude from their field of critical reception all those who are not professional scientists, our amateur artists.

14 *Coal in* Menhir. Instalación 0 *by Menhir, 2015*

The artistic duo Menhir (Coco Moya, Asturias, 1981 & Iván Cebrián, Cuenca, 1980) proposes direct interaction between the spectator and coal from Asturian mountains, in the north of Spain, through digital sound and audiovisual technologies as a way of reconnecting with nature. The spectator–now the performers of the piece–uses their hands to activate the internal sound landscape of the coal rocks through an Arduino[10] system that emits sounds. The work is set up as a material-sound and audiovisual space–or score–that transports whoever performs it to the hidden landscape of carboniferous Asturias through a mental and tactile, in some ways hallucinatory experience, in the words of the artists. Coal acts as a ready-made conductor between the hidden geological strata of mining territory and the social, economic, and environmental experiences of this complex cultural landscape, today extinct. The artists base this work on geomancy, according to which the planet is surrounded by Hartmann lines of energetic force, creating a terrestrial magnetic field. Under the theory of geo-acupuncture, man is called on to intervene with large stones or menhirs in the points where said energetic lines join so as to balance energy. The artists, inspired by these theories, implement a site-specific, non-colonizing, ephemeral audio intervention, which, after disappearing into the landscape, resurfaces and materializes in the exhibition hall through this installation-score.

15 *Minerals in* La Mesure Minérale *by Fabien Giraud & Raphaël Siboni, 2012*

In the film *La Mesure Minérale* (web), shot in a natural history museum, Giraud (France, 1980) and Siboni (France, 1981), Giraud and Siboni investigate the artistic possibilities of technological advances in the language of film

10 Arduino is an open-source hardware platform, based on a board with a microcontroller and a development environment (open-source software), and designed to facilitate the use of electronics in multidisciplinary projects that aim to build digital devices that interact with real-world objects.

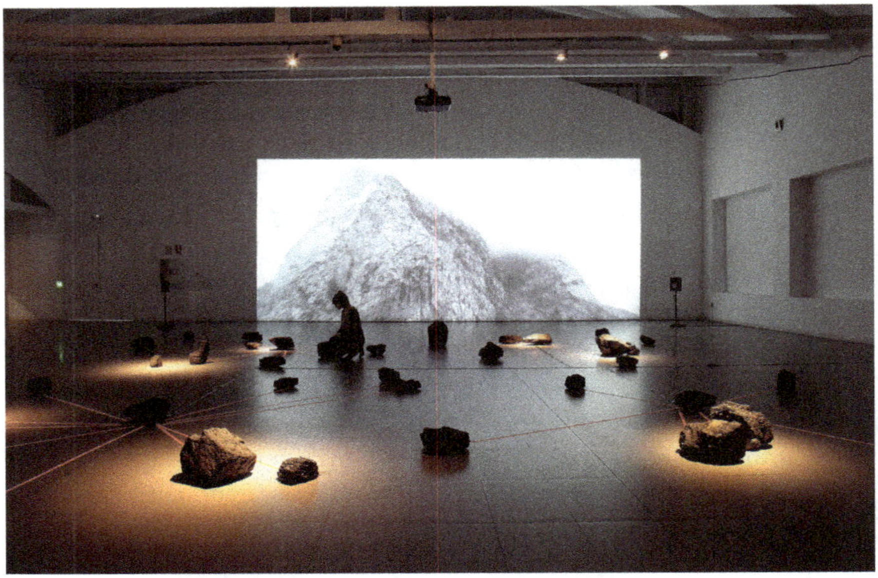

FIGURE 8.2 Image from the installation *Menhir. Instalación 0* by Menhir, 2015.

and photography in the representation of time (human and geological) and its relations of scale beyond humans' capacities for measurement. The work was filmed with an ultra-high-speed camera in an empty mineralogy department closed to the public during its restoration. The museum is presented to us as a mysterious, somewhat sinister setting, immersed in a sound track full of magnetism and tension. Giraud and Siboni present the museum institution as a cultural space of history and memory at a standstill, slowing down the time of its geological materials, a place apart from time as experienced by human beings. The museum is a sort of stratified mineral object that is indifferent to our presence, thus emphasizing the insignificant historical presence of human beings on Earth within the immense scale of geological time. *La Mesure Minérale* presents, in a disruptive fashion, the temporal progression of human beings compared to the frozen materialized time of the minerals, presented as musealized artifacts. It is a spatial-temporal view of the world based on our own inability for measurement.

16 *Halite in* Zechstein Sea *by Carlos Irijalba, 2014–17*

Over several years, the artist Carlos Irijalba (Pamplona, 1979) developed this project on the peculiarities of the Netherlands in terms of the country's proportion of water to land. *Zeichstein Sea* (web) is the result of the artist's collaboration with the company Akzo Nobel, which specializes in mining halite in Hengelo, in which they drilled the strata to remove several 30-meter-long geological remains from a 300-million-year-old dry sea located beneath the Netherlands, Germany, and parts of Denmark and England. That sea is now a solid block of salt buried 400 meters below the surface. Along with these remains of ancestral halite, the installation includes spontaneous maps that are created by spilling seawater directly onto the gallery floor. The work references the materialization of geological time in relation to chronological time on a human scale. Through these geology methodologies, the artist represents the interstices of the Dutch landscape as a hybrid space that is neither solid nor liquid. The channels, ports, locks–all of these aquatic spaces that are somewhere between natural and artificial–represent the Netherlands in all of its hybridity and complexity: Between skepticism with regard to the anthropogenic technical world, but open to admitting humanity's technology as something intrinsic to itself.

17 The Artist as Amateur

As a final analysis of the works presented here, it could be said that contemporary artists interested in the natural space have been driven to work, on the one hand, with a new world view that arose as a result of the new techno-scientific knowledge generated worldwide by the latest research into climate change, the "technosphere" or the Anthropocene (maritime genetics, the economy of coal, atmospheric chemistry, etc.); and, on the other, with disciplines that are closer to the arts, like the humanities or social sciences, but in their updated version–the new environmental humanities, the new cultural and social studies. By putting this fusion into practice, the artists aim to establish cultural and scientific knowledge based on a post-humanist paradigm, which distances itself from the hegemony of human beings and, as a result, considers other non-humans (Latour, *Políticas pagina*) and the displacement of the modern subject. These efforts aim to break with the historic duality between science and culture, returning to and reevaluating a time when human beings and the surrounding space were one and the same. Classical philosophy's divide between subject and object must be revised so that the production of the world and knowledge will be different, to seek another model for relations between human beings and the planet.

I applaud the existence of these post-anthropocentric artistic proposals as models of thought that are alternatives to the until-now hegemonic humanism, even if they are utopian or naive in achieving their aims. By presenting and questioning new paradigms based on primitive matter, these artists inspire in the conscience of contemporary society motivations, empathies, and methodologies to generate human beings that are more critical towards our own nature and towards the place that we inhabit along with other living things and inert matter. Twenty-first century art as an amateur, yet imaginative, proposal is not intimidated by the powerful scientific and technological means and innovations that affirm that they are the only ones capable of resolving this systemic crisis because it knows that not only is a radical change required in the construction of materialities, but also that a change in the subjectivity of mentalities is absolutely necessary, and in that respect art still has a lot more work to do.

Works Cited

Benjamin, Walter. 1934. *Iluminaciones III.* Taurus, 1975.

Biemann, Ursula. "SUBATLANTIC." *Geobodies*, 2015, www.geobodies.org/art-and-videos/subatlantic. Accessed June 2018.

Braidotti, Rosi. *Lo posthumano.* Gedisa, 2013.

Castells, Manuel. *Local y global. La gestión de las ciudades en la era de la información.* Taurus, 1997.

Castoriadis, Cornelius. *Los dominios del hombre: las encrucijadas del laberinto.* 1986. Gedisa, 2005.

Castro Nogueira, Luis A. *La risa del espacio. El imaginario espacio-temporal en la cultura contemporánea: una reflexión sociológica.* Tecnos, 1997.

Dean, Tacita. "Tacita Dean | Selected Works." *Marian Goodman Gallery*, 2018, www.mariangoodman.com/artists/tacita-dean.

de Miguel, Regina. "Artists - Regina De Miguel." *Maisterravalbuena*, http://maisterravalbuena.com/artista/regina-de-miguel/. Accessed June 2018.

Didi-Huberman, George. *Cuando las imágenes toman posición.* A. Machado Libros, 2008.

Echeverría, Javier. *La revolución tecnocientífica.* Fondo de Cultura Económica, 2003.

Eliasson, Olafur. "Ice Watch Paris." *Ice Watch Paris*, 2015, olafureliasson.net. Accessed June 2018.

Engberg, Anna et al., editors. *Studio Olafur Eliasson: An Encyclopedia.* Taschen, 2012.

Fluxá, Bárbara. "Obras & Proyectos." *Bárbara Fluxá*, 2017, www.barbarafluxa.com/obras. Accessed June 2018.

Foster, Hal. 1996. *El retorno de lo real. La vanguardia a finales de siglo.* Akal, 1996.

Giraud, Fabien, y Raphaël Siboni. «La Mesure Minéral.» *Theunmanned*, 2012, www.theunmanned.com/#go-3-2. Accessed June 2018.

"GLOBALE: Reset Modernity!" *Center for Art and Media Karlsruhe*, 2016, zkm.de/en/exhibition/2016/04/globale-reset-modernity. Accessed June 2018.

Gray, John. *Perros de paja: reflexiones sobre los humanos y otros animales*. Paidós Ibérica, 2008.

Guattari, Félix. *Las tres ecologías*. Pre-Textos, 1990.

Halperin, Ilana. "Mineral Bodies: Are We Autobiographical Trace Fossils." *Geologicnotes*, geologicnotes.wordpress.com/are-we-autobiographical-trace-fossils/. Accessed June 2018.

Harari, Noah Yuval. *Homo Deus. Breve historia del mañana*. Debate, 2015.

Haraway, Donna J. "Antropoceno, Capitaloceno, Plantacionoceno, Chthuluceno: generando relaciones de parentesco." *Ileca: Revista Latinoamericana de Estudios Críticos de Animales*, vol. 1, 2016, pp. 15–26.

IPCC. Intergovernmental Panel on Climate Change, www.ipcc.ch/home_languages_main_spanish.shtml. Accessed June 2018.

Irijalba, Carlos. "Zechstein Sea." *Carlos Irijalba*, www.carlosirijalba.com/zechstein-sea/#text-high-tides. Accessed June 2018.

Jarman, Ruth, and Joe Gerhardt. "Earthworks | Semiconductor." *Semiconductor Films*, 2016, semiconductorfilms.com/art/earthworks/, Accessed June 2018.

Kaku, Michio. *Hiperespacio: una odisea científica a través de universos paralelos, distorsiones del tiempo y la décima dimensión*. Crítica, 1984.

Klingan, Katrin, editor. *Textures of the Anthropocene: Grain, Vapor, Ray: Manual*. MIT Press, 2014.

Latour, Bruno, editor. *Reset Modernity*. MIT Press & ZKM. 2016.

Latour, Bruno, editor. *Nunca fuimos modernos. Ensayo de antropología simétrica*. Siglo Veintiuno, 2007.

Latour, Bruno, editor. *Políticas de la naturaleza: por una democracia de las ciencias*. RBA, 1999.

Lipovetsky, Gilles. *Los tiempos hipermodernos*. Anagrama, 2006.

Manacorda, Francesco, editor. *Radical Nature: Art and Architecture for a Changing Planet 1969–2009*. Barbican Art Gallery and Koenig Books, 2009.

McLuhan, Marshall. *Guerra y paz en la aldea global*. Planeta Agostini, 1985.

McNeill, John R. *Algo nuevo bajo el sol. Historia medioambiental del mundo en el siglo XX*. Alianza, 2003.

"Medium Earth." *The Otolith Group*, www.otolithgroup.org/index.php?m=project&id=152. Accessed June 2018.

Mignolo, Walter. *Historias locales, diseños globales: colonialidad, conocimientos subalternos y pensamiento fronterizo*. Akal, 2003.

Moore, Jason W., editor. *Anthropocene or Capitalocene? Nature, History and the Crisis of Capitalism*. PM Press, 2016.

Morton, Timothy. *Ecology without Nature. Rethinking Environmental Aesthetics.* Harvard UP, 2007.

Moya, Coco, and Iván Cebrián. "Instalación 0." *Menhir*, 2015, www.menhir.es/. Accessed June 2018.

Prigogine, Ilya and Stengers, Isabelle. *La nueva alianza: metamorfosis de la ciencia.* Alianza, 1994.

Serres, Michel. *El contrato natural.* Pre-Textos, 1990.

Smithson, Robert. "A Sedimentation of the Mind: Earth Projects," *Artforum*, September, 1968.

Vanden Eynde, Maarten. "Copper Country." *Maarten Vanden Eynde*, 2016, www.maartenvandeneynde.com/?rd_project=764&lang=en. Accessed June 2018.

Vanden Eynde, Maarten. "Triangular Trade." *Maarten Vanden Eynde*, 2017, www.maartenvandeneynde.com/?rd_research=721&lang=. Accessed June 2018.

PART 3

Inspiring Change

CHAPTER 9

Sense of Place as an Enhancer of Empathy and Ecological Consciousness in the Baix Llobregat

Carma Casulá

Abstract

This chapter focuses on the documentary and artistic value of landscape and environment and their representation in contemporary photography. It looks at the impact it has on the social conception of territory and of sustainable practices given its great potential to create empathy and to promote responsibility in human-environment interactions. Attention is paid to the project BxLLt as a reflection of the territory of El Baix Llobregat (Spain) and of how the experience of the sense of place in a specific territory acts as an intensifier of affection that facilitates the raising of awareness about local landscape. El Baix Llobregat has a peri-urban, agro-rural, agro-industrial and industrial character that extends from Martorell—headquarters of the automobile Spanish manufacturer SEAT—up to the Prat beaches adjacent to the international airport and port area of Barcelona.

Keywords

Landscape, photography, empathy, sense of place, environmental awareness, connectivity, map creation.

1 Introduction

The Lower Llobregat (*El Baix Llobregat / BxLlt* in Catalan) represents an invitation for us to take a closer look at how the individual interacts with his or her environment, and to interpret and reflect on people's links with their landscape, which could well be seen as a direct portrait of their society, as well as a part of their identity. In doing so, it is interesting and rewarding to capture how individuals appreciate and cherish their environment or their land, the experience of their perception of it, and how they recognise

© KONINKLIJKE BRILL NV, LEIDEN, 2022 | DOI:10.1163/9789004501270_011

themselves in it; that feeling of belonging to a place with which one identifies, whether a private space or one shared by society: "my place." We are investigating people's idea of sense of place, of rooting, as an intensifier of affection which we propose as a means of enhancing people's ecological conscience. The phrase "to put down roots" means to settle down, to be rooted in a place, to establish oneself permanently in a place, developing connections to people and things. It has to do with the concept of empathising, of *appreciating*, seen as the act of perceiving and ascribing value, which presupposes that we value most what we best perceive or know. The present investigation has focused on the Lower Llobregat area of the province of Barcelona, hereinafter BxLlt.

Any assessment that we make of our environment is the result of the interaction of both objective and subjective factors, as implied by any act of human perception that is not purely receptive, but rather that constitutes a process of identification. We feel the urge to know better that which awakens in us some kind of affection or fondness, and even more so if its identity is enhanced by the stories and feelings it inspires.

All of the above is underpinned by the concept of landscape proclaimed by the European Landscape Convention which defines it as "an area, as perceived by people, whose character is the result of the action and interaction of natural and/or human factors" (Article 1a, Council of Europe 2000). It is the visible face of the culture that inhabits it and with which it creates interrelations. The land has been manipulated and recreated by man/womankind according to his/her needs and he/she projects his/her desires and emotions, his/her history and his/her culture in it. We are all constructors of our landscapes to a greater or lesser extent and, in turn, we are interconnected with them because man/womankind influences and is influenced by the landscape. In sum, we are agents of landscape transformation and at the same time sensitive to the emotions that this provokes in us from which it can be concluded that "we are the landscape" because the landscape is "the cultural projection of a society in a given territory and not only in relation to its material dimension, but also a spiritual, symbolic and identity dimension" (Nogué and Font 397–398; translation my own).

Conflict arises chiefly when we, as individuals, act as if we were detached from the environment in which we live. We must break this disproportionate disaffection, the predominant anthropocentric tendency in current socio-cultural, economic and political thought that leads us to be marooned in the land, disconnected from the environment and the rest of nature, as if it only existed in order to meet our needs as individuals, and as a society. In this respect, Jane Goodall in her well-known 2007 TED talk raised the following question:

> (...) what are we doing to our planet? You know, the famous scientist, E. O. Wilson said that if every person on this planet attains the standard of living of the average European or American, we need three new planets (...) here we are, arguably the most intelligent being that's ever walked planet Earth (...) destroying the only home we have. The indigenous people around the world, before they made a major decision, used to sit around and ask themselves, "How does this decision affect our people seven generations ahead?" Today, major decisions (...) are often based on, "How will this affect the next shareholders' meeting?"

We need to seek strategies to awaken the "ecological self" posited from the environmental philosophy movement by the Norwegian philosopher Arne Naess, who in 1992 established Experiential Deep Ecology as a tool to respond to the accelerated decrease of biological diversity and progressive deterioration of the planetary environment. We must reject the criterion of "man in the environment" in favour of another more relational criterion, "man is the environment," and live in harmony with our surroundings, with our landscape that is our home, for which we have a great duty, such that the rest of the planet's inhabitants benefit from it.

We need to find or intensify ways to extend and promulgate this self-recognition as belonging and interconnected to the natural environment and to the other living beings with whom we cohabit, and as a part of the biosphere on which we depend. This is about activating bio-centric attitudes of "we are

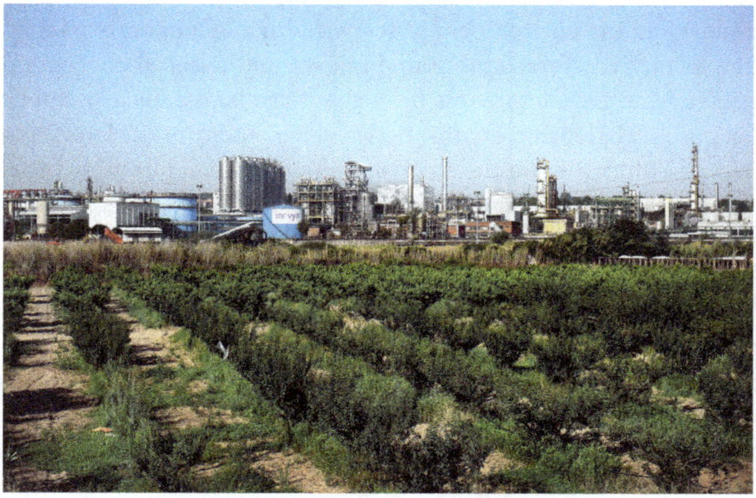

FIGURE 9.1 Martorell and the binomial of its agricultural and industrial landscape, with the headquarters of the automobile colossus SEAT.

nature" and understanding that it is only a matter of change of scale. We are always in a place within another place and we move from one to another.

In this connection, empathy plays an important role, acting as a radar of the brain that helps us to decrypt information from, and emotions of, others, to put ourselves in the place of the other and of other forms of life, by giving us the ability to perceive, share and understand what is felt in a shared or similar context. This is one of the most important human capacities and it is key for our argument because it invites reflection as well as combatting disaffection and the detached use of our environment, so that it is understood more as a place for community building rather than a public space.

It is important that the landscape should be known, interpreted and understood from various standpoints whether geographical, biological, environmental, historical, cultural, sociological or ethical, but also perceived from a more intangible perspective such as aesthetics and poetry. Given this, the capacity of art and artistic attitudes is clearly an effective device in the understanding of our environment and in the transformation of the ways of acting and interacting in it given its great potential to create empathy. It can play a role with regard to socio-environmental problems by promoting the need for a change in the value system in favour of environmental awareness towards more sustainable societies. This interpretation is supported by "a growing tendency to define works of art as patterns of thought rather than as visual acts" (Grande 13; translation my own).

This research is based on the power and influence of the image since we are aware of the value of art as a symbolic language and the artistic significance of the landscape and the environment as scenarios in contemporary photography—given its capacity for synthesis and impact on the social conception of the land. Hence, to raise awareness both about the landscape as a public good and about the relationship existing between the quality of the landscape and the quality of life, we value the contribution of photography as an artistic practice and an effective vehicle to promote sustainability, conservation of land and the development of ecological consciousness given its very strong empathic capacity. It is a valuable tool for visualisation and expression, apt to stimulate reflection and dissemination in relation to the duty of individuals and local communities to become involved in the environment, in an attempt to move towards more favourable attitudes without delegating all responsibility to institutions and to science.

It is important to promote artistic practices with formal and aesthetic parameters based on rootedness, on sense of place, as an enhancer of ecological awareness through transdisciplinary activities, both by individuals and also by collaborative citizen participation, that allow a transformation of values and attitudes.

In this research, the area of the BxLlt has been chosen as site for artistic exploration because it embodies a good part of the contemporary conflicts that affect the environment. One can speak of a landscape cannibalised by a style of growth based on the use of nature as a tool to feed industry and property speculation. It is characterised by degraded areas, a high density of infrastructure, and a big ecological footprint, while at the same time housing a great agricultural tradition now recognised as the "foodbowl service" for the city given its "near-urban farming's importance to the city" (Thomas). In fact, the town of Sant Boi de Llobregat received the Sustainable City Award in Spain in 2017. In short, it is a great scenario with many disconnected actors leading us to create synergies as a survival strategy.

Since one of the aims of this chapter is to draw attention to landscape as public good where the relationship between its quality and the quality of life are expressed, two central courses of action, based on the production of photographic series, have been followed to make the situation visible and to create new maps: physical, affection and connectivity maps. Each of them becomes a unit of artistic and documentary research and production, touching the hearts of passers-by with elements in which they can project their own experience.

The idea is to offer, on the one hand, a personal text on tangible aspects that make up a BxLlt Physical Map as analysis and reflection of the land. And on the other hand, a more collective view of intangible aspects with new texts derived from listening to and working with the local people who reside or pass through the place and with the existing collaborative structures designed to weave an Affection Map and a Connectivity Map, which together with other strategies can transform or recompose the collective imagination. Map creation can generate benefits and allow people to be less stressed, and happier and more connected, leading us to understand our emotional ties and empathy as important factors for the protection of the land. In order to achieve the abovementioned goals, it is necessary to explain first the nature and characteristics of the land used as case study and its various underlying components.

2 El Baix Llobregat

El Baix Llobregat, the official name of the Lower Llobregat, is a complex region located south of the province of Barcelona. It is bounded by three cultural, touristic and landscape icons, namely the Montserrat Mountain, the

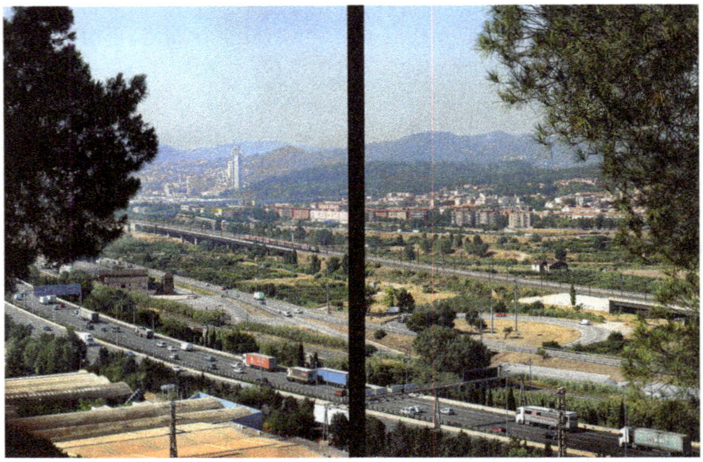

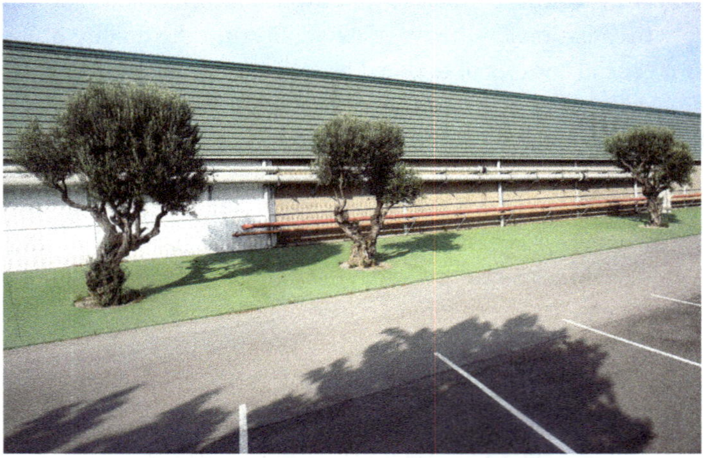

FIGURE 9.2–9.3

Mediterranean Sea and the city of Barcelona with its tourist beaches of fine sand adjoining the El Prat International Airport and the large state port, with dense commercial and tourist traffic as one of the main stops of the Mediterranean cruise lines. It is also, therefore, an area with a considerable ecological footprint.

The Baix Llobregat owes its name to the last section of the Llobregat River flanked by the mountain ranges of Montserrat and Garraf and by the Llobregat Delta Nature Reserve. It is a densely populated area[1] with 486 km^2 of area, with

1 According to the Idescat, the Statistical Institute of Catalonia, the Baix Lobregat had a population of 813,996 inhabitants in 2017.

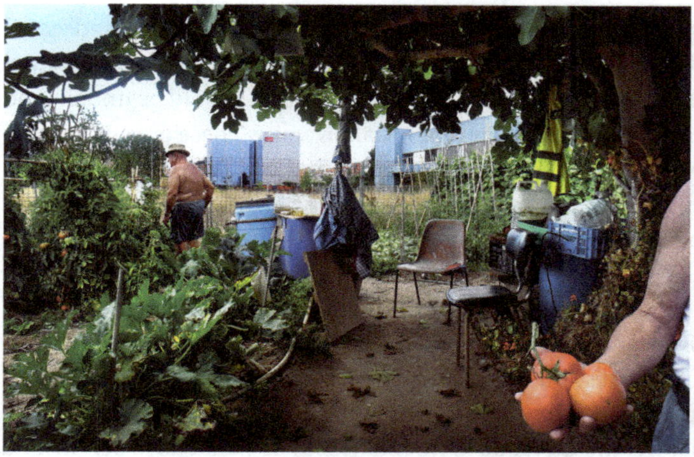

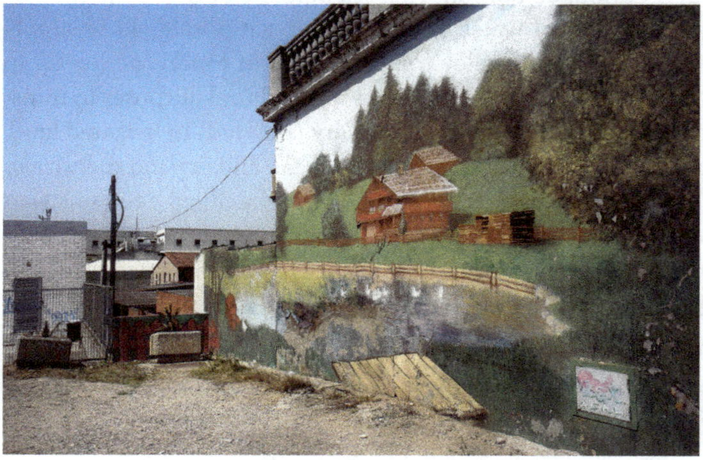

FIGURE 9.4–9.5

the capital of the region being Sant Feliu de Llobregat, with very significant populations such as El Prat de Llobregat or Casteldefels that are part of the first ring of the Metropolitan Area of Barcelona. It is a land where quasi-urban, agro-rural, agro-industrial and industrial land-use is intermixed, and sometimes coexists, with a high rate of activity markedly affected by metropolitan and socioeconomic dynamics and the sphere of influence thereof given its proximity to the big city. It is, therefore, highly politicised ground, due to its potential value and significant planning pressures.

There are numerous elements that act as interferences and physical and visual barriers, creating an impassable wall preventing appreciation and interest in it as well as relations with it on the part of its inhabitants and also users who pass through it. Enclaves abound that fall within the "Third Landscape" category coined by Gilles Clément covering areas that man abandons and makes useless, that are left to develop naturally, and that should be understood as spaces of opportunity given their biological diversity. These are spaces generated by the dense networks of infrastructures: regional railways and the AVE high-speed train, the A-2 Motorway, the C-32 main road, secondary roads, the non-motorised vehicles corridor, irrigation water channels and other underground channels not visible but well-marked such as gas and water mains. All of them are barely interconnected and run basically parallel to the course of the Llobregat River with high residual, acoustic and visual pollution, in addition to the resulting incessant air traffic and maritime traffic.

The agriculture and industry of this region both owe their development to a crucial infrastructure: the Canal de la Infanta (1820) or Canal de la Serenísima Infanta Doña Luisa Carlota de Borbón. This is a pipeline of 17,420 km (and 25 km of secondary channels) of water from the Llobregat developed to supply the fields of the left side of the valley from the town of Molins de Rei up to Sants. It was built with private capital for purely agricultural purposes to transform about 2,700 hectares of fertile land from dry (rain fed) to irrigated land, and thus go from a single crop to 3 or 4 crops per year, and expand the variety of products. It quickly became a dynamic axis of the regional economy that far exceeded the forecasts of its developers, causing another unforeseen benefit: the industrialisation of the area in the mid nineteenth century. Faced with such success, the right bank did not want to be left behind and built another channel in 1855 from San Vicente dels Horts to El Prat de Llobregat. Since the Civil War (1939) its area has been steadily reduced to give way to continuous urban growth, and today the only working stretch is the 5 km from Molins de Rei to Cornellá that supplies a community of 280 irrigators on 375 hectares.

This region was one of the first industrialised areas of Spain and a benchmark for the industrialisation process of southern Europe. This displaced and gradually reduced primary agricultural and livestock activity. For good reason it is said that the Llobregat is one of the busiest rivers in Europe.

The socio-economic system of the valley is based on the extraction of raw material that is used in its heavy industries and in the production of cement and textiles, all of them with great need for water, hence their installation in the river course. Besides, one of its hallmarks is the Colonia Güell textile village (1890) designed by Antonio Gaudí, a workers' colony of exceptional importance from an industrial heritage perspective.

Within it there are spaces of great ecological and cultural value combined with others with a significant presence of the primary sector, mainly agriculture, an aspect emphasised after the recognition in 1998 of the Parque Agrario del Baix

Llobregat, BLAP (Baix Llobregat Agricultural Park) that occupies a good part of its territory. This is a protective demarcation whose purpose is to preserve the cultural value and diversity of its landscape to counteract pollution and urban expansion within its 3,500 hectares of agricultural land, and also guarantee the large-scale local supply and diversity of food, greens and fresh vegetables when in season, largely organic, for markets and tourists. This is local food that has obtained the Protected Designation of Origin with the classic advantages of agriculture close to the city, known as its "foodbowl," reducing the ecological footprint of food by minimising transport costs and avoiding a large part of the greenhouse gas emissions related to it; an increasingly significant factor in food security in view of serious challenges such as climate change and oil shortages.

BLAP is a member of the Peri Urban Regions Platform of Europe (PURPLE) and has obtained the Protected Geographical Indication (PGI). It has become a great urban garden in the form of a green belt that takes in not only Barcelona city but also the 14 surrounding towns, improving the quality and freshness of the urban air and interconnecting the citizen with the land and with nature. It is a working space as well as a space for disconnection and leisure that preserves the identity of the place; and, very importantly, it maintains the connectivity of the wildlife habitat since it extends along protected wetlands of the Delta and the Mediterranean coastal vegetation itself.

In fact, in a short piece written by Aaron Thomas for the World Wide Fund for Nature (WWF) the Baix Llobregat Agricultural Park is considered as a great green lung in the middle of the dense metropolitan area of Barcelona that needs to be preserved. But it can hardly be otherwise since its access is complex, as it is literally flanked by infrastructure and fences. Some very good initiatives already exist, among which we can highlight those undertaken by the NGO Desos Opción Solidaria, founded in 2001, with social justice and economic and environmental sustainability in mind. This organisation focuses its efforts in the municipality of Sant Boi de Llobregat where it works to recover abandoned or disused plots of the BLAP and revitalise them through the production and marketing of organic vegetables. Faced with an increase in situations of poverty and inequality arising from the economic crisis, since 2015 the Central Parc cooperative has worked to facilitate insertion into the workforce through training and the implementation of occupational projects in agriculture. The Centre de Estudis Comarcals del Baix Llobregat (CECBLL), founded in 2005, is a public entity that does important and activist work in defence of the land, the dissemination of sustainable themes, as well as its identity, its history and the recovery of historical memory. Also interesting are the activities of the Fundació Sigea, founded in 1994, a non-profit association specialising in environmental education programs and on raising awareness of the need to

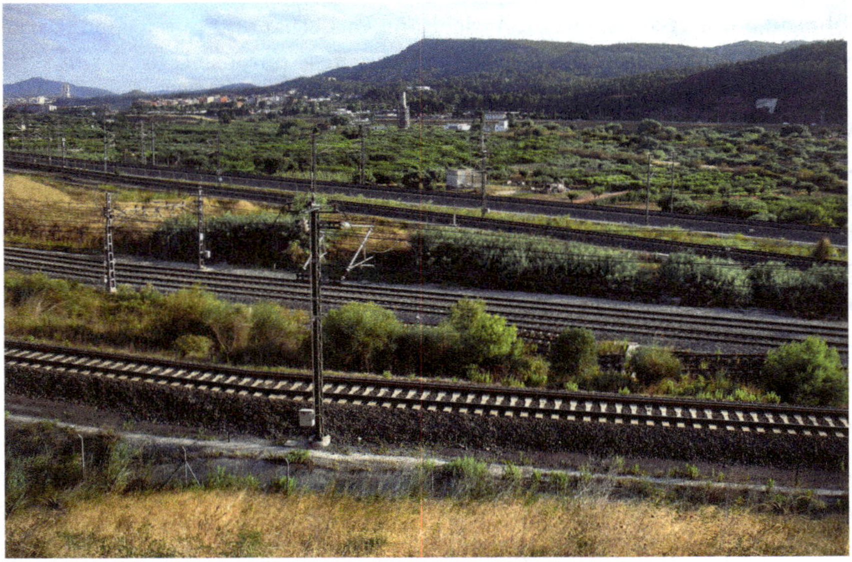

FIGURE 9.6 Market gardens and infrastructure in the LLAP.

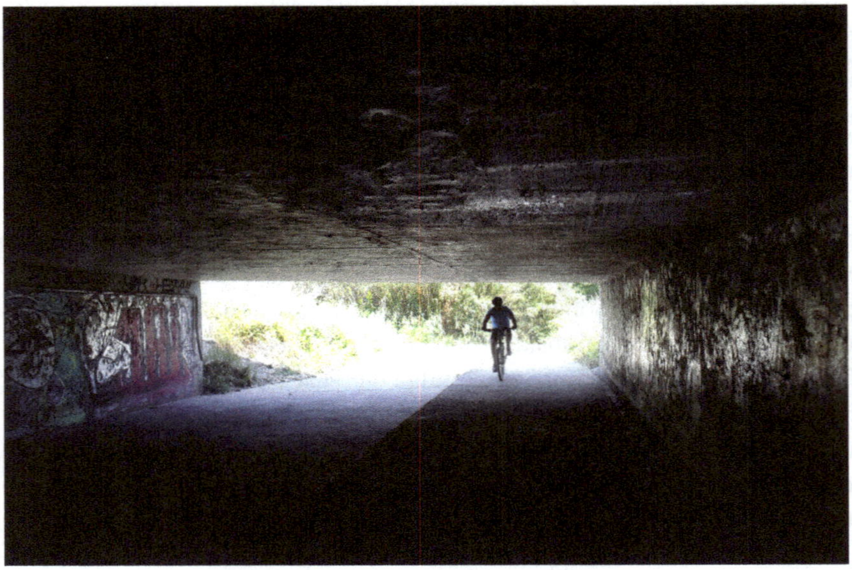

FIGURE 9.7 Passage for animals and pedestrians under infrastructure.

evolve towards more ecological attitudes in everyday life, and the promotion of our heritage. Another example worthy of mention is that of Somos/SomBici, founded in 2014, a non-profit social cooperative that promotes environmental routes with a local farmer.

All of this shapes the identity of the BxLlt, its conflicts, its singularity, its pertinence and interest for the present study.

3 Methodology of the BxLlt Physical Map

The focus of attention reflected in the Lower Llobregat Physical Map developed as the basis of this research focuses on "the green spot" that includes spaces with agricultural, ornamental, leisure and autochthonous vegetation left wild or managed in this area of work.

To draw up the BxLlt Physical Map, the socio-environmental context has been studied. Special attention has been paid to its problems and the consequences derived from them such as environmental, agrarian and urban quality, de-ruralisation, non-transitability, non-visibility and aesthetics of the territory. Photography, understood as a vector of reflection and change, has been used to develop a visual reading of the area.

Much interest has been placed on valuing the opinions of the people who inhabit it, to establish dialogues about their appreciation of the environment and connection with the place. This is not only in relation to coexistence and interaction with land qualities and problems, but also in relation to how the inhabitants perceive this landscape, be it public, community or private. The vision of this external environment is laid out "from their window," whether they live in farms, apartments, *pairal* houses—ancestral local homes—or new-build blocks and, also, the changes made to humanise them with their orchards or gardens. We inquired about aspects such as legacy, beauty, chaos, others, ours, mine, etc.; aesthetics and emotions about their spaces, in short.

We noticed that there is a remarkable disaffection and distancing between individuals and their landscape, perceived almost exclusively as a "place of transit" both by passers-by and by its users and residents. This is basically caused by complex mobility, due largely to the proliferation of dense areas because of concentration of industry, homes and numerous infrastructures that cross it, in addition to geographical elements. Access is difficult and there is little connectivity between intermediate spaces, many are isolated and others could even be considered non-places.

In addition to taking a look at planning, environmental and sustainable practices and policies, and the institutions and professional teams that exercise them, as well as to assess the level of incidence of fields such as environmental humanities in the public arena, it has been essential to approach neighbourhood groups, new residents, farmers, users of the place and existing collaborating entities and associations—those who can be the best guardians of the land.

Thus, work has been carried out with the Observatorio de la Urbanización (Planning Observatory) of the Autonomous University of Barcelona UAB, the Centre de Estudis Comarcals del Baix Llobregat, Coboi (Public Social Innovation Lab in San Boi de Llobregat), Fundesplai (Foundation of the Esplais)[2] and the Observatorio del Paisaje de Cataluña (Landscape Observatory of Catalonia).

The first step in the type of field work presented here was to establish contact with farmers. Accessing their market gardens was a very trying task, but once they understood the aim of this study, the farmers became strategic reporters, completely immersed in the weaving of the Connectivity Map in addition to being participants in the Affection Map.

Following the successful reception of the project *Monsanto no es santo de mi devoción* (Casulá 356) by both farmers in other Spanish locations as well as the artistic circles, a similar kind of approach as an experiential experience, was used when first contacting the inhabitants of the BxLLt. This project aims at creating a Memory Bank of Farmers in Spain in order to support and document traditional modes of living from and cultivating the land. These are mostly initiatives where the farmers collect self-produced seeds that are later stored in containers that preserve the best of them, but also the most productive, or the most cherished. The result is a compilation of visual and written narrations about their origins and why they chose them. And thus, in the same manner that there are seed banks, as a consequence of the ecological alarm bells, this project also intends to preserve a system of life and culture based around the seed. This involves working on themes such as local products, the population ties established around the seeds, their specificity and localism, as well as the relations built around them. As in each agricultural area, a photographic series has been woven around a selection of thirteen seeds because there are thirteen annual lunar months by which farmers are basically governed. Here are some stories:

Pere, artichokes (46): "That is the best smell in the world, it smells like home, it smells of damp earth, it is the one that we all like. In El Prat we have always had the artichoke, it is the typical crop. We cut the plants leaving only a handspan and then out (of the ground) we split them into several pieces, we sold them for cuttings. Before we bought the Tudela plant (Navarra), several of us agreed to it and the trucks came from there, but now we produce it ourselves

2 Non-profit entity with leisure time clubs whose mission is to educate children and youth, strengthen educational entities and the Third Sector, improve the environment and promote citizenship and social inclusion.

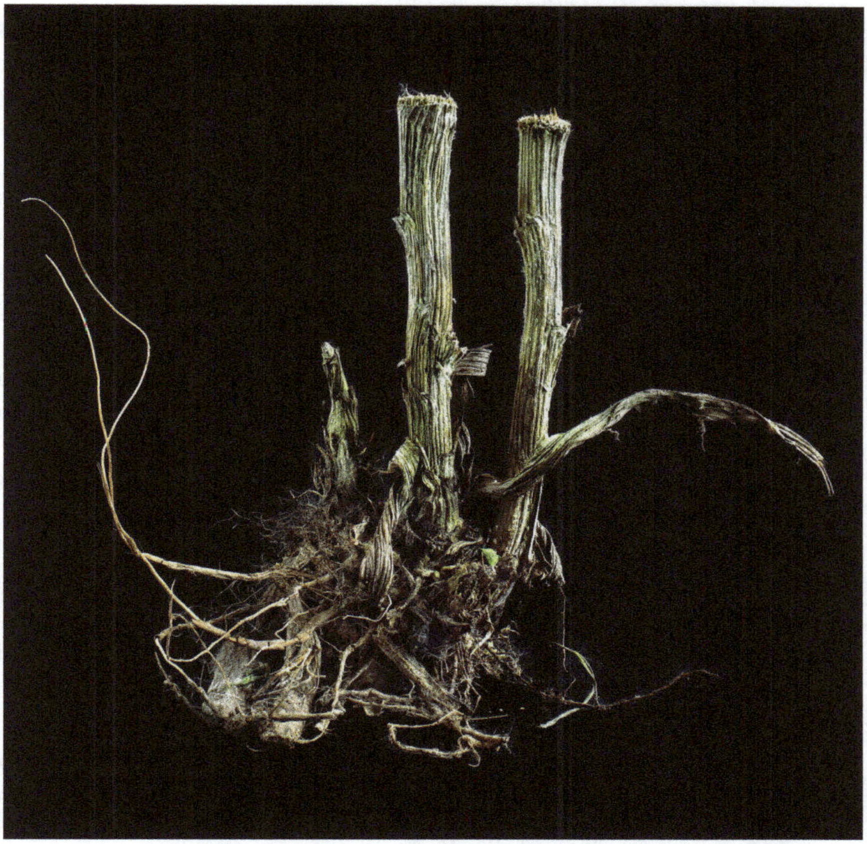

FIGURE 9.8 Pere, *artichokes* (46).

and we live from it all year, from the artichoke and plant harvest, although it is a lot of work and I have trouble finding workers. The typical one is the white one of Tudela, although the Italians like the purple spike, in France the round one. Everything is for export, I sell very little for domestic consumption."

Yolanda, muchamiel *beans* (41): "These beans are from the Abastos Fair of Jaén, near the Military Hospital. My grandfather was a fair worker and my grandmother had a sweet stand. For eighteen years, we talked about the war—Civil War—and the post-war period; they ate only the husk of the beans and potatoes because they sold the rest. In my house, the bean is like a cradle, a fundamental part of my life. They are from Jaén, so to cook them, the oil and beans go in, and sweet onion over a medium heat for an hour and a half, and then with a fried egg. It was typical to cook this dish on Saturday, on holidays. This is what I ate during the 20 years that I lived in Mallorca with my grandmother."

Given that one of the most effective ways to become aware of a place and its characteristics is precisely to establish emotional and communicational links with its environment and its ecosystems, an active and positive approach has been chosen as a strategy in afflicted areas such as the BxLlt, often less intrusive than a negative one. This project involves the creation of an Affection Map and a Connectivity Map to create a network of protection through closeness, a web of emotions concerning those places preferred by their inhabitants, those towards which they have a greater feeling or emotional bond. Those places that they would wish to preserve, those that they would value or, simply, those that they appreciate for their beauty or characteristics, in sum, what they love and care for. In sum, the idea is to make visible a network of people's affection to place and literally, put in these maps how their situatedness works as an enhancer of environmental awareness.

In order to dissolve the sense of a barrier that stops much of the contact between its inhabitants and between the individual and the place, a cartography has been woven that summarises the possibilities of connection and physical closeness existing in this territory given its scarce signage and the high degree of difficulty involved in its perception. Such orientation pursues to emphasise what unites people with their roots, with their places of life and origin through the artistic language of photography, so that a transformation of the collective imagination and a weaving of a network of roots are enacted. This can generate a change of perception that increases empathy with that environment. It can also facilitate the recognition of the specialness of each place with the consequent feeling of greater satisfaction with oneself and with one's surrounding and, as a result, the feeling of belonging to a larger community of neighbours. It is an exercise that unites past and present, and implicitly the future, that touches the hearts of passers-by with elements in which to project their own experience.

This leads us to cartography, the science that studies and represents the experiences in the land with the drafting of geographical maps. In this sense, according to the architect Francesco Careri, defender of the "practice of walking," cartography is the record of the succession of stages, the drawing of a path resulting from the accumulation of experiences, which is given as a legacy to one's offspring. And with regard to the idea of the path, he mentions that "[t]he only architecture in the Palaeolithic period was the path, the first anthropic sign capable of imposing an artificial order on the territories of natural chaos. Space (...) slowly began to transform into rational and geometric space generated by the abstraction of thought" (Careri 50). Thus, lines in the land can be read as lines of thought that modulate and attempt to manage the landscape that is filled with connotations to which it is somehow linked.

Anne Whiston Spirn, an American landscape architect, artist and renowned professor of Landscape Architecture and Planning at the MIT, argues that the language of landscape is our mother tongue because "humans touched, saw, heard, smelled, liked, lived and modelled landscapes before the species had words to describe everything they could do. Landscapes were the first human texts (...). The trees were recognised as guides towards food and water; the chirping of birds, as warning of predators" (Spirn np; translation my own). She affirms that talking and reading the landscape is a by-product of living and a survival strategy. This is an organic vision that she implements as the head of the West Philadelphia Landscape Project (WPLP). There she places schools and landscape education for children as the cornerstone of community development and fosters the "art of citizenship" making children know their own place and how to take care of it, of their landscape.

Some antecedents, that may seem antagonistic, converge in our proposal of maps of affection and roots as a cumulative experience of the territory. Worthy of note are, for example, the approach of the world maps owned by Louis XIV, King of France. These were a valuable gift from his chancellor Cardinal D'Estrées, who entrusted their drafting to the prestigious cosmographer of the Serenissima Republic of Venice Friar Vincenzo Coronelli (Venice, 1650–1718). He made two globes of 11 metres in diameter and 2.3 tons of weight per piece: one, the Celestial Globe, representing the sky as it was the day of the birth of the Sun King; the other, the Terrestrial Globe, encompassing all the cartography known of the Earth at the time. In doing so, Friar Coronelli did not begin from direct experience because he never left his cell, but rather he resorted to the stories of explorers, sellers, missionaries, adventurers, merchants, soldiers or sailors who came to his monk's cubicle to entrust him with their observations and experiences after their travels through these worlds of God. This means that his maps were the result of subjective impressions that he historicised in the form of paintings and inscriptions of all the landscapes, figures and human actions known until then.

At that time, the Assembly of the Royal Academy of London determined that Coronelli's globes were the most correct and the most replete with astronomical and geographical novelties of all the other globes hitherto printed, and this was backed up by the Physical-Mathematical Academy of Rome in 1692. Today these globes are located in the Parisian Bibliotheque National de France. Vincenzo Coronelli founded in 1684 the first international geographical society that he called Accademia Geografica degli Argonauti, based in Venice.

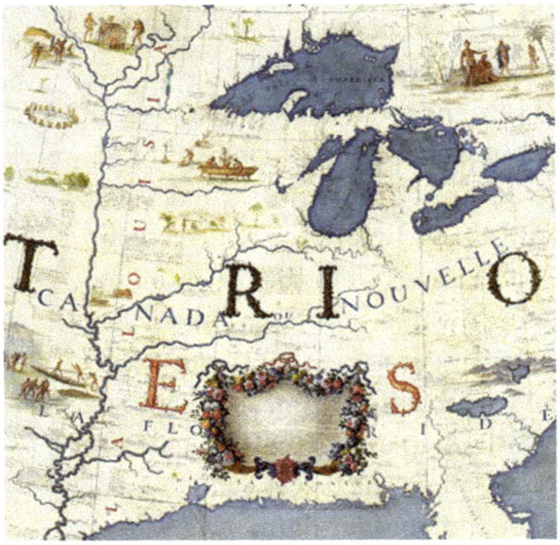

FIGURE 9.9 Detail of the Terrestrial Globe of Ventian Friar Vicenzo Coronelli.

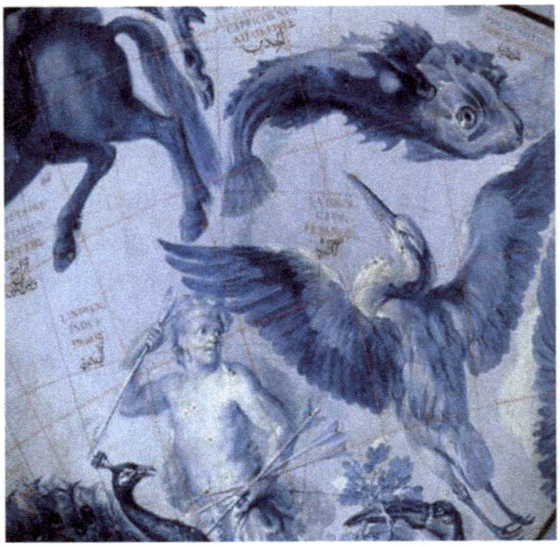

FIGURE 9.10 Detail of the Celestial Globe of Vicenzo Coronelli.

4 Building Maps in the Twenty-First Century: Cartographies of Emotions and Cartographies of Affect

Although cartography is a science and, as such, a branch of knowledge considered accurate based on the materialisation and rationalisation of the experience acquired by man during his travels with methodical observation, these appreciations show that the conception of reality does not remain constant since our origins but rather it is changeable, thus it can indeed be said that we are children of our time. The abovementioned globes were considered the most complete and advanced maps until the eighteenth century—the GPS (Global Positioning System) of that time—but with the eyes of today these cartographies seem more like fables.

Knowledge in the twenty-first century benefits from greater specialisation. Mechanical advances have changed our way of perceiving things, our logic and our criteria of credibility, although, even today, much of what is deemed to be accurate has never been seen by the human eye *per se*, but through photographs, macro-computer screens and lenses which, as such, distort the observed subject albeit only a few microns; the *minimum error* contemplated by science.

For many artists and thinkers, as in the case of Bill Viola, the use of technology does not distance us from man, it is only a form of his expression and an extension of the highest rational capacities of the individual. For them, spirituality is linked to technology and artificial intelligence is seen as an accumulation of human knowledge. For Viola, nature is not in opposition to technique, rather it works as a primaeval technology, capable of causing alterations in human sensory experience.

In line with this, Google Maps and Google Earth map application servers have become vital tools in reading the territory and drawing up our maps. They represent a further agglutination of human thought as a result of technology; a new landscape where the mental and the physical, where the real and imaginary become intermixed. These are sophisticated applications that, at the same time, request their users to contribute with their photographs of the places visited, and as databases or search engines they value very positively the information of their users.

Currently, the ability to orient oneself has decreased considerably in those people who habitually resort to GPS to locate or find a place. This is due to lack of attention to detail and that is the reason why in this article I would like to reclaim the use of memory and its evocation as a way to empathise with places and relate to them. Perception does not mean a pure reception, but rather it constitutes a process of identification, and in this, memory is a primordial factor. It requires learning and reasoning since, given the complex variety and deluge of information to which we are all subjected, our senses and our brain work through selection. Images seen and information received previously condition our way of seeing and perceiving and codifying what is being perceived in the now. And that is what is proposed when weaving Maps of Affection, to capture memories and offer them to others on the great schematic offered by Google Earth as a medium to compose a photograph with which to cure disaffection, in order to better treat the environment as a place for community building, and transform attitudes towards a more empathetic and organic, ecological society.

In addition to using photography as a creative and differentiating product, it may be convenient to incorporate some of the criteria contemplated by applied psychology in emotional marketing and in advertising in order to increase efficiency and thus, reach a wider audience and inoculate this ecological awareness as salvation. The ability to create or awaken empathy in the audience is the key to successful communication. A clear example is the approach promoted by Digital Menta, as an agency specialising in the sector of digital marketing, is to move the citizen to action, by presenting reality as something that can be embraced and changed for the better. Creative and attractive content is no longer sufficient in the face of so many impacts that bombard us in practically any place and in any activity that we engage in. The excess of information makes us invulnerable to it. This current trend is aligned in the creation of sentimental connections with clients to be able to generate empathy in the receiver through emotionally charged stories that move them to action. "Make yourself happy" is the message, and not just the buying and selling of a product. Campaigns are designed to provoke something, evoke a certain memory or awaken a feeling that can later be related to the brand. The secret to a good communication is not so much what is being told, but rather how it is being told, the personalising of problems by putting faces to them, explaining them and their consequences on a certain individual with which the receiver can feel identified. This coincides with the criteria proposed to configure the Maps of Affection.

A highly valued tool in structuring objectives in emotional marketing is to weave an empathy map that defines the receiver as a person and not as a segment of the population. Its construction revolves around the four following concepts FSHS (FEEL, SEE, HEAR, SAY) which will provide the answer to four sets of questions: 1. What does the customer think and feel? What matters most to him/her? What motivates him/her? What are his/her dreams? What obstacles stand between him/her and his/her desires? What are his/her problems? 2. What does he/she see? What is his/her environment? Who are the key people around him/her? What kind of issues impact him/her daily? 3. What does he/she hear in his/her environment? In what kind of medium? What do the people in his/her inner circle tell him/her? Which people have an influence on him/her and how? 4. What does he/she say and do? What attitude does he/she have? Who does he/she usually talk to? Does he/she thinks he/she has influence on someone? Are there differences between what he/she says and what he/she thinks? In our environmental context these are very important questions for overcoming the imbalance in the relationship between the individual and his/her environment: the biosphere.

To dissolve the sense of a barrier that stops many contacts with place and between inhabitants, a Connectivity Map has been woven that groups together the possibilities of connection. In order to do so the contributions provided by the residents of the BxLlt have been gathered together with information collected through other tools such as Google Earth, Google Maps and various maps of the area as well as other sources such as the Observatory of the Landscape of Catalonia that offers free of charge on its website cartographies with pedestrian and road itineraries, vantage points and other information of great interest. But our proposal, by necessity, is to find and establish where the gaps are in this land, the space in which we live, and in which we can introduce ourselves: the *void*. Thus mapping these empty places or passing through this land that is so densely built and criss-crossed by solid lines becomes a challenge: "it is the memorandum of a succession of stages (...). The cartographic map, although it is static, presupposes the idea of narration, is conceived in terms of an itinerary, it is an odyssey" (Calvino 89; translation my own).

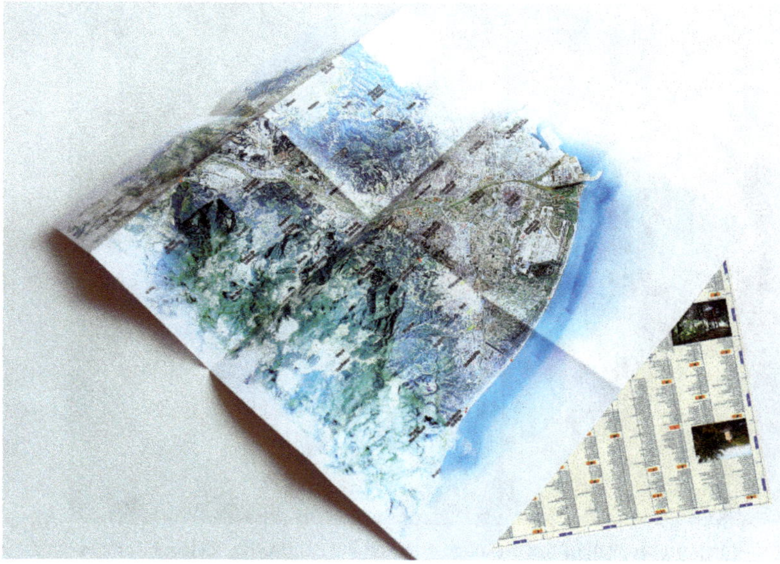

FIGURE 9.11 Prototype of publication of the BxLLt Affection Map and Connectivity Map.

Another line of work has produced an Affection Map charged with revealing experiences that show the degree of empathy the dwellers have with their land and thus recover stories about memory and the transformation of the area of the BxLlt. To compose the map and capture these memories and emotions, I have either collected photographs taken by the protagonists or else photographed their favourite places in the area and also what is remembered as the most precious for their loved ones, adding the written accounts of these memories and thus threading together at least two enclaves per person to weave a complementary experiential cartography drawn up by encounters and empathy with the place. To demonstrate this interconnection, as a medium, a frame captured from Google Maps has been created where the photographs have been signposted and incorporated, whilst the narratives have been included into the lateral panels and back as is the custom in traditional, tourism or road geographical maps. This facilitates the creation of a path through which to break down the disaffection and to treat the environment as a place of community. In sum, to transform attitudes towards a more empathetic and organic, ecological society.

Among the contributors we have:

Montse (*49*): "My favourite place is here, on the terrace, with this view over the fields with the cranes of the port in the background, the planes passing by and the boats that look like they are on top of the vegetation. I can get a glance

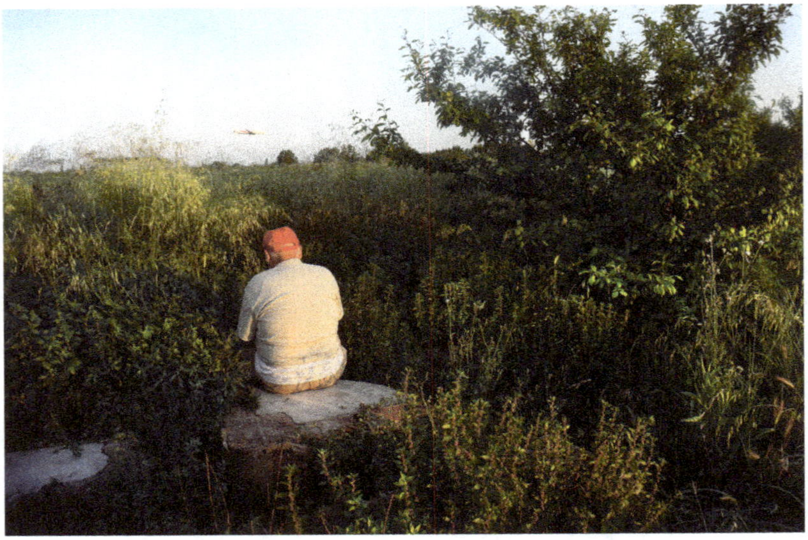

FIGURE 9.12 Pep on the wall of the drainage ditch where his brother Cisco, Montse's father, used to sit.

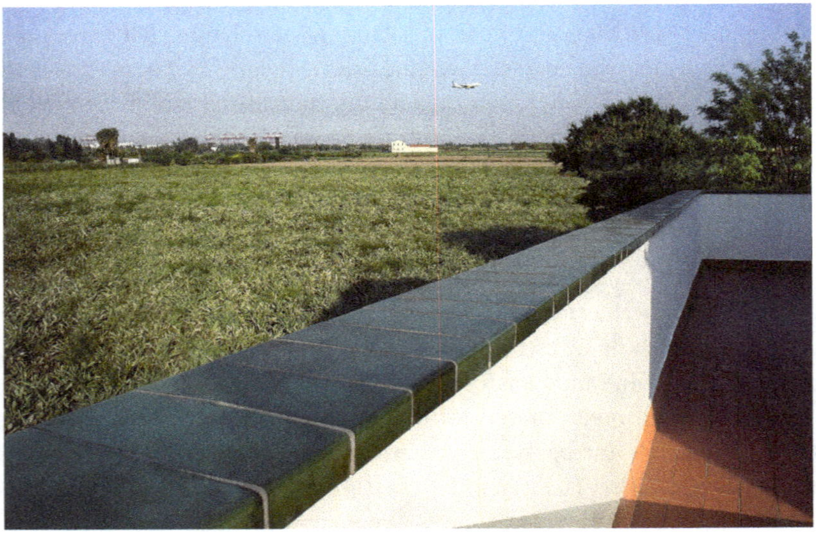

FIGURE 9.13 The terrace of Montse's home among the fields of Prat.

of other places, you can see everything from here. I love to relax after work, quietly, having a drink without sitting or anything, just looking. The curious thing is that it is the same point that my partner Alberto (57) likes. He lived for many years in that nearby house, in an adjoining field, and I used to cycle on this road but we had never seen each other. We did not know each other until a few years ago. There is a place within our land where every time I go past I remember my father Cisco (Francisco); he liked to sit on the wall of this water channel and daydream while he looked at his fields towards the sea, as now does my uncle Pep, his younger brother who still lives with us as he did all his life. They worked side by side and they were together all the time."

Jordi(17): "I feel like I'm coming home, that I've already arrived when I see the hermitage of San Ramón. It is even better from the plane, although many times I jog up, you can see everything for 360 degrees from the sea to Montserrat, the Sagrada Familia, the beach. It is one of the best places in Barcelona. Since parents got married in the crypt of the Gaudi Colonia Gúell, whenever there is something to celebrate or someone comes, we go there so that they can see it; and every time they tell us that they kissed under the windows that look like butterflies and not at the altar. It is a very precious place for my family."

Also, as a complementary strategy not reflected in the maps but of informative value, the numerous participants of the study have been asked to photograph the places where they do the (supposed) differentiated collection of organic waste, glass, paper, plastic and metal packaging in their homes. This has been done to make the interlocutors reflect on the responsibility for and care of the environment with personal acts related to ecology, sustainability and recycling in their most private and immediate space such as their home. Various attitudes have been uncovered from this: personal pride in exercising responsibility, consensual irresponsibility because, even though they know the need to implement more sustainable practices and the obligation to comply with the regulations, they fail to do so out of laziness, lack of space and time, because of misinformation or because they consider its direct impact to be of little importance, or simply because the eco collection points are located too far away from their home.

In the Connectivity Map, the locations of collective and institutional recycling areas and facilities have been included, both in the agricultural, industrial and urban sectors, as well as the peripheral eco collection points and the compost bins. All these stand out as efforts made to promote the 5 R's: Reduce, Repair, Reuse, Recycle and Regularise.

The majority of people who have taken part in this project have very much enjoyed to find out about the enclaves where something positive about the environment can be done.

The participants have been very cooperative and even enthusiastic about the Affection Map and about the fact that their most personal sites will be taken into account. With regard to the knowledge of those enclaves of affection they prefer or that could reflect the preferences of their loved ones, around 70% did indeed show that they were aware of this personal aspect of

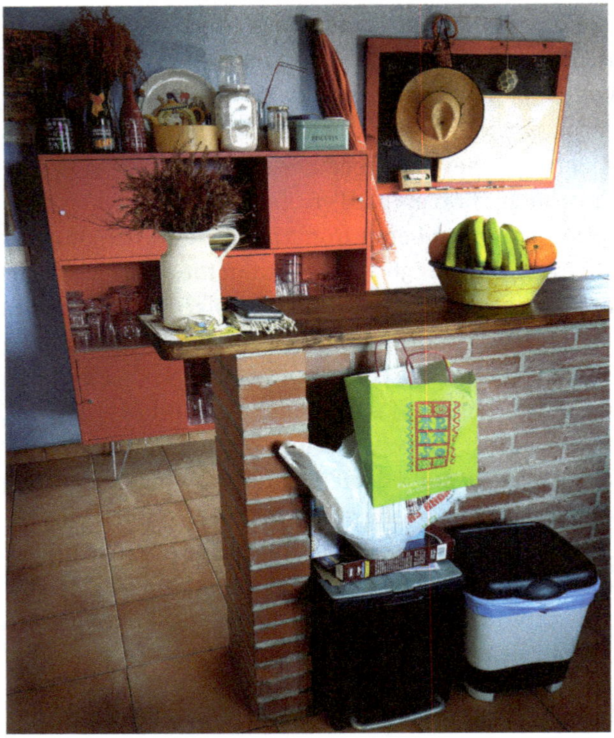

FIGURE 9.14

their close surroundings. However, a considerable part of them showed reluctance and even discomfort in relation to the proposal to provide photographs that reflect their private recycling areas; only 32% of the participants did so, which shows a certain guilty feeling and recognition of their unethical conduct. On the other hand, and in order to count on the participation of children and young people, the Fundesplai Foundation (an integrative association for culture and leisure time) has been contacted to access a younger population among whom there is a notable presence of immigration. Young people should be taken into account for exerting great influence on their elders, especially if they happily adopt a more ecological approach to life. Of the photographs that have been contributed to the present study by the children of the Esplai, there are abundant images of refuse and recycling areas inside their houses with a participation of 78%, much higher than the adult community.[3]

3 BOE translates as the Spanish Official State Gazette.

At the same time, like any problematic area, the BxLLt shows contrasting realities. For example, the BOE of 8 June 1994 published guidelines for the planning of several large-scale infrastructures that affected the Llobregat Delta, such as the high-speed train works, the new Logistics Activities Zone of the Port of Barcelona or the extension of the El Prat de Llobregat airport with the third runway—the latter on its own involving the cutting of about 15,000 trees in the protected area. In stark contrast to this, in the course of the present investigation, the population of Sant Boi de Llobregat has received the 2017 Sustainable City Prize awarded by the Fundación Fórum Ambiental with the support of the Spanish Ministry of Agriculture and Ecoembes. For this reason, local institutions have shown an interest in the strategies proposed in the present investigation since they believe they can be implemented as a tactic to reduce tension and project fluidity and affection towards the territory. These Affection Maps and Connectivity Maps produce benefits and through them work is done to increase the degree of well-being and connection between its people and between them and their landscape. People have to be made to feel "at home" and to perceive the environment as an extension of their abodes, as a community space and not only as a public space. Emotional ties and empathy need to be strengthened as important factors of territorial protection. These new mappings aim to reduce costs and time, as well as minimize less ethical behaviors by revealing physical connections that are difficult to access and emotional links that are sometimes not easily perceived unless unearthed through the citizens' reflective appraisal of their environment.

Local authorities of the Baix Llobregat consortium have shown interest in the publication and distribution among residents of these maps, and in the creation of a blog associated to that of CoBoi Lab (Public Social Innovation Lab in Sant Boi the Llobregat). As a complementary activity, photography workshops and field trips have been planned to promote contact with the landscape and all its actors. The Baix Llobregat Physical Map will be published in book format by the Planning Observatory of the Autonomous University of Barcelona (UAB).

Finally, the realisation of this project shows that we must break with the tradition of art as merely decoration or the "cherry on the top"; a work with artistic sensitivity can be found anywhere and an artistic attitude can work without a work of art. It is important to seek the integration of artists in interdisciplinary teams as thinkers and executors of plans and projects, to promote artistic attitudes and more sensitive and aesthetic actions and to understand space as a work of art, where collaborations with architects, landscapers and engineers are also possible. In sum, there is a need to establish constructive dialogues among the disciplines of art, aesthetics, ecology, engineering and culture.

Works Cited

Calvino, Italo. "El viandante en el mapa." *Colección de Arena*. Siruela, 1998.

Careri, Francesco. *Walkscapes. Walking as Aesthetic Practice*. 2002. Culicidae, 2017.

Casulá, Carma. "Al Natural_PRS. Parque Regional Sureste de la Comunidad de Madrid." *Arte y Ecología*, edited by Tonia Raquejo Grado, UNED, 2015, pp. 343–62.

Clément, Gilles. *Manifiesto del Tercer Paisaje*. Gustavo Gili, 2007.

CoBoi Lab. www.coboilab.cat/ca/. Accessed 7 May 2021.

Digital Menta. "Cómo crear un Mapa de Empatía para conocer a tu cliente." *DigitalMenta*, 17 Aug. 2015, www.digitalmenta.com/blog/como-crear-un-mapa-de-empatia-para-conocer-a-tu-cliente/. Accessed 17 June 2018.

European Landscape Convention ETS No. 176. Florence, Council of Europe, 20 Oct. 2000, rm.coe.int/1680080621. Accessed 9 July 2018.

Goodall, Jane. "How humans and animals can live together." *YouTube*, uploaded by TED Global, June 2007, www.ted.com/talks/jane_goodall_how_humans_and_animals_can_live_together.

Grande, John K. *Diálogos Arte-Naturaleza*. Lanzarote, Fundación Cesar Manrique, 2005.

Idescat: Statistical Yearbook of Catalonia. "Altitude, Surface Area and Population. Municipalities." www.idescat.cat/pub/?id=aec&n=925&lang=es. Accessed 11 Sep. 2018.

Naess, Arne. "The Three Great Movements." *The Trumpeter: Journal of Ecosophy* vol.9, no. 2, 1992.

Nogué i Font, Joan and Pere Sala. "Los catálogos del paisaje." *Gestión del paisaje. Manual de protección, gestión y coordinación del paisaje*, edited by Jaume Busquets Fábregas and Albert Cortina, Ariel, 2009, pp. 397–426.

Spirn, Anne W. "Ser uno con la naturaleza: paisaje, lenguaje, empatía e imaginación." *Boletín CF+ S 38/39*, 1 March 2006, habitat.aq.upm.es/boletin/n38/aaspi.html. Accessed 6 May 2021.

Thomas, Aaron. "Barcelona Market Gardens: Harvesting Sustainability at the Huertas." 1 March 2012 wwf.panda.org/wwf_news/?204379/Barcelona-market-gardens . Accessed 4 June 2018.

CHAPTER 10

Building Stories to Change the World

Interview with Starhawk

Carmen Flys-Junquera and Beatriz Lindo-Mañas

Abstract

This interview was conducted at the University of Alcalá in Madrid, Spain in 2018 within the context of an international conference focusing on the power of stories, myths and arts to envision change. Although Starhawk is known primarily as an activist and ecofeminist, her presence as a plenary speaker was based on her role as novelist of a contemporary ecotopian society. Her ecotopia is strongly based on scientific knowledge, particularly aspects of renewable energies and permaculture, as well as in the embracing of diversity in all its manifestations, spirituality, rituals and stories. In her novels the imagination and the arts become essential to inspire the people to create their ecotopia and resist the invasion of the hegemonic militaristic government. During the conference she also gave a workshop centred on the imaginative process of reinventing better and more sustainable worlds through stories. This interview precisely highlights the imaginative process in action and how the imagination can inspire change.

Keywords

Imagination, fiction, ecotopias, sustainability.

Starhawk (1951--) is best known as an activist and an ecofeminist. She founded *Earth Activist Training*, an organization for teaching permaculture design grounded in spirituality and is a veteran of many progressive political movements working against war, nuclear weapons, anti-globalization and international solidarity movements. She is also a major voice in earth-based spirituality, Goddess religion and Wicca, being a co-founder of *Reclaiming*, an activist branch of modern paganism. Most of her thirteen books and her international seminars revolve around these topics. As such, she has been

© KONINKLIJKE BRILL NV, LEIDEN, 2022 | DOI:10.1163/9789004501270_012

interviewed frequently.[1] However, her invitation as a plenary lecturer to the I International Conference in Spain on Environmental Humanities: "Stories, Myths and Arts to Envision Change", was based on her role as a novelist. Perhaps this role has been less visible, but she is the author of three novels, the ecotopian *The Fifth Sacred Thing* (1993), its sequel, *City of Refuge* (2016) as well as its prequel, *Walking to Mercury* (1997).[2] During the conference she also gave a workshop, focusing on how to reinvent worlds through stories. Given the contexts of the conference and the related research projects[3] this interview is centred on the novels and the role of stories to envision a change.

The interview was conducted at the University of Alcalá, Spain on July 5th, 2018.

---

1 Interviews with Maya Lilly do deal with the novels. However, most interviews feature on webs of related organizations and deal with Starhawk's facets as pagan, earth activist, etc. See Berta, Blumberg, Doyle, Mackey, McKay, and Rubin. The only in-depth academic interview we have found is that of Rigoglioso.

2 In order to facilitate understanding of the interview, major aspects of the novel are summarized. In *The Fifth Sacred Thing*, as a result of climate change, California has suffered a severe drought and coastal areas (including L.A. and San Francisco) have been flooded. The social unrest causes the government to fall and a militaristic group, the Stewards, supported by fundamentalist Christians, takes over. The citizens of the San Francisco area rebel and create a utopian community, privileging the production of food over weapons, education, health, ecology, the arts and diversity. The novel carefully traces and describes the social system and the ensuing debates. Governance is carried out by representative councils and consensus. The economy is based on working hours banks except teachers, artists and healers who are considered essential and unable to limit their work to countable hours and receive a set stipend. Education is applied (such as science) and includes training for consensus and negotiation. The community has chosen non-violence and the Defense Council is composed of nine old women (given their age, experience and wisdom). Healing combines Western and Chinese medicine as well as other herbal remedies and alternatives. The society honors diversity of languages, cultures, familial and sexual arrangements, religions and so forth. On the contrary, southern California suffers the Stewards regime, one of strict moral "purities" and hierarchies, police control and privatization of all resources where women are forced to return to traditional gender roles. The Stewards decide to try to invade the San Francisco area again and the utopian community's commitment to non-violence is put to test. In the sequel, *City of Refuge*, the utopian community realizes that the Stewards will keep trying to invade them and the only solution is to oust them from power by getting the southern communities to rebel. Much of the novel is based on the efforts of lead characters Madrone (healer) and Bird (musician) to inspire change in the oppressed south and teach them how to regain self-governance and build a utopian community.

3 "Environmental Humanities. Strategies for Ecological Empathy and the Transition towards Sustainable Societies; Subproject 2: "Stories for Change" HAR2015-67472-C2-2-R (MINECO/FEDER) and "Research Activities in Cultural Mythcriticism" Ref: S2015/HUM-3362 (CAM/FSE).

I: *Why do you write fiction?*

STARHAWK: I really love the process of writing fiction and creating another world—that work of creating characters and seeing them going through their stories and their changes. I often asked myself that same question [laugh]. And I like to read fiction.

I: *What do you think is the power of fiction in contrast to nonfiction essays, science, journalism, academic writing or the so-called serious writing, given that fiction is often not considered serious?*

S: There is power in fiction, certainly in speculative fiction. Ursula LeGuin has often spoken about that. But for me I think there is a lot of power in writing fiction of any sort because it is emotional, it brings people into the experience of another human being in a way that you identify with it, you experience it yourself vicariously. You feel the feelings of another and that has a lot of power. One of our problems in the world right now is that there's a massive empathy deficit and fiction is a good antidote to that. Speculative fiction can do a number of things; it can warn, it can inspire, it can be a thought experiment of taking an idea and exploring where it might go or how it might actually work in reality.

I: *And the power of imagination? Ursula LeGuin often wrote about the power of the imagination as the only way to free ourselves from the constrictions of the present* (45). *The imagination also figures very prominently in your novels; for example, the fact that Maya and the Defense Council say that the major war is one against the imagination, that if our imagination is limited, then there is an inevitable defeat* (FST 238). *Saying that any war is first fought in the imagination, certainly doesn't sound very rational, scientific or even realistic.*

S: Right. Well in fiction you're released from rational logic. In science you are supposed to document everything and quantify and record and make a rational argument for what you're doing. And in fiction you are released from that; you can create whatever you want. It's also a lot less expensive than trying to create the alternatives in real life. That is freeing to the imagination. Where our imagination goes, then reality can follow. If we can't actually envision a better world, then, how are we going to create it? When we have no model, we have no place to direct our energy to create it.

I: *And with the imagination we can accept magic. David Abram speaks about magic being a simple change of consciousness* (9).

S: The definition I've always liked is Dion Fortune's "the art of changing consciousness to the will" (qtd. in Pizza and Lewis 82) which to me is very much like that. It's when we shift our consciousness that interesting things can happen.

I: *To what extent do you think the humanities in general, including fiction, fine arts and so forth, can contribute to environmental consciousness and to a more just sustainable world?*

S: The humanities are really vital to dealing with climate change, because obviously science alone is not enough. Science is pretty clear about climate change, but the issue with climate change is not actually the science. It's not even about knowing when and what we need to do, because that's also pretty clear. The problem is the human factor of why is it we can't grasp it and do something about it. Those problems are social and political and cultural, and so we have to use our cultural and social tools to address them.

Climate change goes against our underlying cultural narrative or story of progress. Whether we think of ourselves as religious or atheist or whatever, we believe that, somehow or other, science has made things better and it's going to continue making things better and the world is getting better and better; and climate change says "Well, no, actually it's not, right?" And this very progress in technology is actually what's killing the world. In order to challenge that we have to have some kind of a positive counter narrative. We can't just have a narrative that says the world is actually dying and we're doomed because people just go into denial. We need to be able to figure out how we tell a powerful story that says: "Yes, we are facing immense danger because of our choices and our way of life. But we also have the capacity to turn this around, and to create a world that is more just and more balanced and more beautiful than the world we have today."

I: *Do you think we have a chance of creating that? Do you truly believe that there is still hope?*

S: Yes. I think we will be contending with climate change regardless of what we do at this point, and we will not get back the world that we had in the past, but I think we still have the potential to create a viable, liveable world that is fair and just. If you look at climate change as a massive ecosystem degradation, we have the knowledge and the skills to massively regenerate ecosystems on a global scale. If we do that, I believe the climate will come back into a kind of a balance that we can live with, in a world that generates abundance for everyone. But it won't be a world that generates enormous profits for the same people that are reaping them right now, and that's what we're up against.

I: *You're saying that we have the science and the ability to do so, but how do we achieve the will to so do. Other eco-fictions seem to propose doing so by force (as in Karen Traviss'* Wess'har *series [2004–2008], where the aliens impose ecological balance on other planets or in Matthew Glass'* Ultimatum *[2009], where politicians accept negotiating the change only as a consequence of a few nuclear bombs). In your novel you also give us a ruthless military group supported by fundamentalists, as the democratic government collapses due to, among other things, climate change. You wrote that in 1993. More than two decades have passed, but the combination still rings true. How do you see this in our current society?*

S: Well, it's alarming, actually. I did write that in 1993 and today, in 2018, we have Donald Trump doing his best to impose a kind of inept irresponsible chaotic despotism supported by the fundamentalists and evangelicals, even though he exemplifies absolutely everything they say that they hate. That was a part of the underlying premise in *The Fifth Sacred Thing*: these people who called themselves Christians but were not actually Christians. There are Christians in the book, like Marie the nun who, to me, embodies the best of Christianity. And I did that deliberately because I didn't want it to be an anti-Christian book and I was teaching in Christian college while I was writing. But, in the novel they use the terms of Christianity to basically cover an agenda that's authoritarian and patriarchal. And, yes, I feel we are much closer to the negative parts of *The Fifth Sacred Thing* today than we were back when I was writing in the late 80s and early 90s.

I: *Which is why I was wondering about that hope. Do you really believe a peaceful revolution, as in* The Fifth Sacred Thing *(and, albeit, to a lesser extent in* City of Refuge*) is possible?*

S: Well, I would say there is hope. I also teach permaculture design and I'm trying to keep up with the agroecology of the science around that. You know there's a huge move now to look at soil as the place to draw down carbon because the world's soils have a huge carbon debt from agriculture, farming and erosion. Soil can heal again, when we adopt the practices that build soil organic carbon, which is humus, the stuff that gives soil fertility. A huge amount of the excess carbon in the atmosphere actually doesn't just come from fossil fuels. It also comes from erosion and the depletion of humus from chemical and industrial agriculture. There's a researcher, Dr. Rattan Lal at the Soil Carbon Institute in the University of Ohio, who says we have as much excess carbon in the atmosphere from the Dust Bowl in the 1930s as from every automobile ever invented.

So to me, the hopeful things are the movements that people are making. We need to look at the way we produce food. We can actually produce food without tearing up or plowing up the soil, without having to do all these things that are so destructive and that cause so much damage and we can do it in ways that actually regenerate soil fertility. These methods take carbon out of the atmosphere and store it safely in the soil, where it increases nutrients and helps repair the water cycle. Humus holds more humidity in the soil which gives you more resilience to drought and helps prevent flooding. Building soil organic carbon has no downside—only multiple benefits. And there are people all over the world that are working on doing that. California actually has state policies and grants, the Healthy Soils Program, to encourage and support farmers to switch to regenerative methods. So, again the knowledge is there, the tools are there, the things that we need to do around climate change are actually nice things. We don't have to go out and sacrifice every third infant or something; we need to grow food in healthy regenerative ways, we need more people on the land doing work that's restorative and regenerative rather than sitting in offices manipulating figures in finance. That part is all very, very hopeful. The question is how do we get from here to there, when we have massive power interests in the way and that part means the strategy and the activism.

I: *I'd like to continue a little bit with the science. As you were talking, I was imagining the character Cress, while digging trenches, designing the rain catchment and run-off diversions and swales for the future* (City of Refuge). *In your novels as in any futuristic fiction, there are several scientific advancements. What are they based on? To what extent are they simply imagination or extrapolations of current science? For example, the crystal powered computers…*

S: The crystal computers are based on my imagination. In *The Fifth Sacred Thing*, I was writing in the late 80s, early 90s. As I mentioned this morning, my first computer was an early model without even a hard drive. I had it when I was writing *Truth or Dare: Encounters with Power, Authority, and Mystery* (1988). It was this little thing, with floppy disks and it kept crashing whenever I had too much in a chapter. And then I think I finally got a hard drive for *The Fifth Sacred Thing* and I remember my "techy" friend saying: "It's a 44-megabyte hard drive, you can put a whole novel in there!" And I was so excited! "Yes!" And now I have single picture files taking more than 44 megabytes. But when I was writing *The Fifth Sacred Thing*, at the very beginning of the era of email and internet, I wanted an excuse to not have to deal with computers in the books, so I made up the crystals. I found out from experience that when I was in a good mood, the computer worked and when I wasn't it didn't work. So, I made the computers based on intelligent crystals, that wouldn't work unless they liked you and shut down in the face of the invaders.

I was also interested in the question of what would science and technology be like if, instead of coming from this framework that says that only what's classifiable is real, and there's no consciousness in this system, what if we had a framework that said everything is imbued with consciousness? If that were true, then maybe our computers and our systems would be different; we'd be working with silicone and with crystals and maybe we'd be communicating with them rather than just manipulating them.

I: *And the* ch'i*? There is the incident with the bracelets where Bird is capable of manipulating the energy in the electronic monitors but also others where Madrone is capable of healing people by shifting the energy.*

S: The *ch'i* comes from Chinese medicine. I wanted to show magical energy at work and Chinese medicine is a technology that works directly with energy. In *The Fifth Sacred Thing* energetic healing works a little bit better.

I: *In your novels you clearly have a utopian goal. But what I find most interesting is the process of building that society. I am particularly curious about your five criteria for judging things: usefulness, beauty, sustainability, nourishing for the earth (or not damaging) and nourishing for the soul. Where do they come from? And, do you think these criteria could really be applied at a community level in our world? Have you tried anything similar?*

S: I made them up. But they came from a lot of thinking and talking about how we make decisions about technology, how do we decide what things we want to use and not use. You know, if we reduce down to a lower energy budget, how do we make choices about what to use, what kind of technology is appropriate? And I think those five could be used to make those decisions in real life situations, and they'd be good criteria.

I: *Do you apply them to a degree?*

S: Yes, I do. I recently went through a period of having to consolidate my office and get rid of a lot of stuff. I had to apply some of those criteria. Is this actually useful to me? Is this something that takes more energy to maintain than it gives back to me? William Morris said: "Have nothing in your houses that you do not know to be useful or, believe to be beautiful" (108).

I: *Communal living is also a frequent aspect of your novels, yet you don't sacrifice the possibility of other permanent personal commitments or different family arrangements or flexible family styles. Do you think that flexibility in our society is viable?*

S: Yes. Diversity was an important value in the books. So rather than saying 'here's the relationships we all should have…' I envisioned a multiplicity of options. The same kind of relationship doesn't work for everybody. Some people really are happy in a monogamous couple. Some people are not. I know people that are quite happy being polyandrous, in multiple combinations. Despite creating those relationships in the book, in my own life they don't work so well, but for other people they do. People can be gay or straight or gender fluid, all sorts of different possibilities, and why do we have to legislate that? What if we actually encouraged everybody to explore and figure out the kind of relationship that is going to work for them?

I: *The governance in the novels is based on the local councils. How effective are they? As your character Sam complains, "and about a thousand hours of meetings"* (FST *18*)*, do you think that could really work in real life?*

S: Yes. To make direct democracy work, it's helpful to have good meeting facilitation and skills and a good decision-making process. That's something alternative communities have slowly been working on over the years. There are many systems. For example, in the permaculture world, a lot of people like a system called sociocracy, which was started by Kees Boeke (1884–1966) who was a Quaker. His disciple, Gerard Endenburg (1933-) inherited an electronics business and decided to run it in a non-hierarchical way. He took Quaker-style consensus process and simplified it, putting it into a business framework and creating a governance structure around it. Many people find his system really helpful. If you're going to work non-hierarchically, you need training and you need to consciously design your governance structure. That's partly why I wrote *The Empowerment Manual: A Guide for Collaborative Groups* (2011) because I saw so many groups foundering on the rocks of conflict. Because we grow up in a hierarchical structure, we're shaped by it; to work in a different structure, people need re-training. In a hierarchy, conflict is often resolved by bringing in a higher authority. If you remove that authority from the system, you need to find another way to move conflict out. Otherwise, there's no one who has the authority to step in and say: stop it! So conflicts can cycle endlessly and the group implodes. You need to have some way to vest that authority in the group consciously if you're not going to have it vested in a person or a role.

I: *Another major issue in the novels is that of healing, both physical and spiritual, and you draw on a large variety of medical knowledge. Right now, in Spain, the new government is made up of qualified experts, many of them scientists, which is a good thing. However, they are all very much against any type of alternative medicine, homeopathy, reiki, and so forth, increasing the already existing debate. So, do you think this type of holistic medicine or alternative medicine is viable? Is it an answer to our problems? What do we do with Western science and the pharmaceuticals? Here, of course, is the main role of the character Madrone as a healer.*

S: Yes, many of the alternatives are viable and valuable. There are things that Western medicine does really well, and there are other things it doesn't do so well. And there are things alternative modes of healing can do. Take acupuncture, for example. Even Western medicine accepts that it works, although nobody can give you a Western explanation as to how. And yet it has become accepted as a valid treatment because in so many cases it does work. And Western medicine doesn't always cure. Somebody has cancer and they go through all the chemotherapy and everything. Well, sometimes it works and sometimes it doesn't. Many people, like those politicians you mentioned and many people on the left who are devoted to rationality are not able to acknowledge that Western science is actually a frame, a paradigm. It is not the full reality. And even within that frame and that paradigm there are things that we can't account for like acupuncture or like dowsing—finding water with hazel sticks or rods. People have been dowsing for thousands of years to find water or find minerals all over the world. Lots of water drilling companies use dowsers. It's one of these things that are done and is accepted by rational people, not just by "woo-woo" spiritual people, but by redneck drilling operators and oil companies. Again, there's no Western framework to explain why that works but it works. So, there's a lot that our Western frame doesn't account for. What I tried to depict in the books was a system that was open to all those modalities and using the best of each. There are times, such as when my husband had a heart attack this spring, that I'm really grateful we had Western medicine to put in a stent and save his life. And that's also a kind of magic. But it's not the only approach to every illness.

I: *And Madrone learning to work with the bee mind?*

S: The bees actually came from a walk I was taking with David Abram.[4] He got a grant in graduate school to travel around in Indonesia and Nepal and meet shamans. He was a sleight of hand magician and wanted to meet them magician to magician. He came back with the idea that shamans weren't so much working with supernatural powers but with natural ones, the animals and the plants that were around. And he and I were taking a hike together down in L.A. At one point, thinking about the book and going up in the hills, we saw some beehives. He said: "Maybe you could do something with the bees." Thinking about it, reading about them I found the bees *are* magic. It's well documented that bees hatch and they immediately go to work, going through different jobs in every phase in their life cycle. They know what to do next and their hive functions as a single hive mind, not as distinct individuals. Again, it's something science has documented but doesn't really have an explanation for. But I started to imagine what could happen if you really became one with the bees.... We do have this tremendous capacity even with our normal human senses to identify chemicals by the smell or the taste. And there are dogs that can smell cancer. So, what if you took that further? What if you could draw on the ways bees communicate through scent and pheromones?

I: *You place a lot of weight on the importance of the arts in the novels, and that's, of course one of the issues of this conference. Making things beautiful, singing, music, and storytelling are some of the major*

4 See Abram (3–29).

aspects of both novels: the arts move things, make things happen. Moreover, even in the economic structure you created, artists, like healers and teachers, are the only professions with standard stipend. This is particularly interesting because artists usually don't have any support in our society.

S: Our society doesn't value pleasure—and that goes back again to a basic worldview that says anything that is actually rooted in the body, in the real world, is of lesser value than these abstract things that are disembodied—spirit, or in our case now, it's more disembodied profit. So, of course, why would art and music be important if pleasure is not? But in the world of *The Fifth Sacred Thing,* because embodiment and interconnectedness are important, then pleasure is sacred: Just a good standard pagan value. And so, things that give pleasure are important. And I believe that if we thought that way... things would change. I've worked a lot with kids who are coming from really poor and deprived backgrounds, inner city kids doing permaculture gardening. And the truth is you can get them to do gardening; but it takes a lot of work and a lot of persuasion and sometimes you need to pay them to do it. But every time we would start doing art they would light up and they get so excited, with that joy of expressing themselves. And they all love music. You couldn't get them to turn their headphones off because they're constantly listening to music and rhythm and they would write music and poetry. And when I think about it, I couldn't tell you anything I learned in my seventh-grade algebra class. I couldn't solve an equation to save my life even though I got an A in it, but I could sing you massive amounts of all the Bob Dylan lyrics I learned at the same time. So, we need people who use music to teach. In ancient times, music was always a core part of education in culture and storytelling and memory because it gives you that capacity to remember things. So, what if instead of forcing people to memorise lists of things, we sang them? What if we actually used that capacity in teaching and in education and made that central, instead of a peripheral thing that gets cut as soon as the funding drops?

I: *You also speak a lot about rituals, and different types of spiritualities, religions in your novels. Do you think rituals and spirituality are essential?*

S: Rituals and spirituality are the things we do communally to reaffirm our collective connection and those bonds help us face the things that are too hard to face on your own. There's no society on Earth that doesn't have some form of ritual and ceremony and spirituality.

Ritual marks what's important, so if things happen and you don't mark them with ritual, it's like saying that it didn't matter. If someone dies and there's no memorial, no funeral, no nothing, that's like saying that life didn't matter. We need those ceremonies to help us make the emotional transition, shifts and changes.

I: *I asked you before about the movie and you said they were working on a TV series, but you don't know when...*

S: No... When we started the crowdfunding, it was never to make the movie. The crowdfunding was to get a package together that we could take out to someone who could make the money. With the crowdfunding we made something like 60 - 70 thousand dollars, which is great. But to make a movie like *The Fifth Sacred Thing* would take tens of millions of dollars in the budget. And the movie industry has changed so much in the last 10 years since we started this project, where they're not making many movies like that. The movies they're making are either very small low budget art films or they're trying to make big blockbuster, Marvel Comics. Like everything else, the film industry is suffering from late-stage capitalism, eating out the middle of it. But TV is having a golden age and there's all kinds of outlets. At this point we have a production company that's interested in making a series and they're putting together the package right now that they'll take out to a distributor, and then they'll go ahead and into development. But the beautiful dream of me and my friends making it and doing it totally the way we wanted to, is not going to happen.

I: *There will have to be concessions there.*

S: Yes, it's going to have to be much more mainstream. But there are people with good values.

I: *And on a personal note, if you don't mind... how much of these novels are projections of your own life? I see elements, I am imagining that you were part of the hippie movement...*

S: *Walking to Mercury*, which is the prequel, I would say, it's not really autobiographical, but it has a lot of my life and sort of pushing to the edge of where I might have gone if I actually made the wildest choices and if I had just run away from home, like I thought about doing, and didn't come back and go to college. So yes, it took parts of things, sometimes it has a seed of something real and then takes it to a different extreme. But I will say that, I, Johanna in the book, had a dear friend, Isis, who I was in high school with.

I: *Who is not the pirate Isis of the novels.*

S: No, not the pirate Isis, but I named the pirate after her, because I knew she'd like that. She is African-American and we were friends in high school and lost touch with each other a while and then came back and were pretty good friends for a long time. And she would read my manuscripts and at one point she told me she had left the manuscript in the car and her girlfriend read it and they had this enormous fight. They were like: "You never told me you and Starhawk were lovers." "We weren't;" and she says: "Yes, you were, here. It's says you were." "No, it's fiction, like don't try to tell me..."

I: *I wanted to ask you about the founding incident of the rebellion in the novel: the nine old ladies "taking the pickaxe to the street." Is that anyway related to Mother's Day protest in Vermont in 1987?*

S: No, not directly. But there is also the "Reclaim the Streets" (RTS) movement near London (1996) (http://rts.gn.apc.org) at a time when they built that gigantic goddess puppet and they were blasting out loud, really loud, like disco music, the sound speaker and then underneath the skirts of the puppet, they had some people with jackhammers and they were jackhammering up the street and planting sunflowers. But I didn't know about that when I was writing the book and I'm not sure they knew about the book when they were doing that demonstration.

I: *Thank you very much. I would like to finish simply asserting that your novels are an inspiration. They illustrate the hope and the work to be done if we are indeed to change and that without this work, that of the arts and the imagination, we are, in effect, tied to the binds of our current framework and unable to transcend them.*

Works Cited

Abram, David. *The Spell of the Sensuous*. Vintage, 1996

Berta, Marilyn. "We Are Sacred as the Earth: An Interview with Earth Activist Starhawk." *Treehugger*, 2018, www.treehugger.com/culture/we-are-sacred-as-the-earth-interview-starhawk.html.

Blumberg, Antonia. "How a 'New World' Would Look to One of the Most Famous Pagans Alive." *Huffpost,* 2016, www.huffingtonpost.com/entry/starhawk-city-of-refuge_us_56ccb6f4e4b0ec6725e42ca2

Doyle, Jim. "EARTH MOTHER / Author Starhawk Is a Spiritual Leader for Bay Area Witches." *Sfgate*, 2018, www.sfgate.com/bayarea/article/EARTH-MOTHER-Author-Starhawk-is-a-spiritual-2678373.php#photo-2151113

Earth Activist Training. earthactivisttraining.org/

Glass, Matthew. *Ultimatum*. Atlantic, 2009.

Le Guin, Ursula. *Dancing at the Edge of the World. Thoughts on Words, Women, Places*. Grove Press, 1989.

Lilly, Maya. "Starhawk Interviews about Her New Novel, *City of Refuge*." *The Fifth Sacred Thing*, 2016, www.thefifthsacredthing.com/2016/02/16/starhawk-interviews-about-her-new-novel-city-of-refuge/.

Lilly, Maya. "What Does Transformation Look Like: A Dialogue with Starhawk." *Reality Sandwich*, 2018, realitysandwich.com/320846/what-does-transformation-look-like-a-dialogue-with-starhawk/

Mackey, Mary. "Interview with a Witch" *Mary Mackey*, 2016, marymackey.com/interview-with-a-witch/

McKay, Dodie G. "Starhawk, Permaculture and *The Fifth Sacred Thing*." *The Wild Hunt*, 2016, wildhunt.org/2016/10/starhawk-permaculture-and-the-fifth-sacred-thing.html.

Morris, William. *Hopes and Fears for Art: Five Lectures in Birmingham, London and Nottingham (1878–1881)*. Ellis & White, 1982.

Pizza, Murphy, and James R. Lewis. *Handbook of Contemporary Paganism*. Brill, 2009.

Reclaiming. reclaiming.org/

Rigoglioso, Marguerite. "Interview with Starhawk." *Feminist Theology: The Journal of the Britain & Ireland School of Feminist Theology*, vol. 13, no. 2, Jan. 2005, pp. 173–83. *EBSCOhost*, doi:10.1177/0966735005051946.

Rubin, Howard J. "Reclaiming the Dark: An Interview with Starhawk." *The Sun Magazine*, 1983, www.thesunmagazine.org/issues/93/reclaiming-the-dark.

Starhawk. *The Fifth Sacred Thing*. Bantam, 1993.

Starhawk. *Walking To Mercury*. Bantam, 1997.

Starhawk. *The Empowerment Manual: A Guide for Collaborative Groups*. New Society Publisher, 2011.

Starhawk. *City Of Refuge*. Califia Press, 2016.

Travis, Karen. *The Wess'har Series*. Harper Collins, 2004–2008.

CHAPTER 11

Eco-interactions: Art and Community

Elena Sánchez-Vizcaíno and Lucía Loren Atienza

Abstract

This article aims to address how the arts can contribute to making or re/creating a more sustainable society. In order to do so, it explores and analyses the work of Lucía Loren and her experience with environmental art. First, this chapter will describe the important role of local communities and how every change is initiated by the community's needs. Following this, it looks at Lucía's work, her process, her interaction with the space and the community, and the short and long-term impact of her interventions/interactions. As an outcome, we will show how Lucia's work can be described as a tool for community members to embrace the arts to re-imagine change and impact their own environment.

Keywords

Social creativity, eco-interactions, community building for change.

1 Introduction

While some politicians are still trying to figure out whether or not our world needs to respond to our current environmental crisis, the humanities are already a step ahead. The origins of environmentalism in the 19th century started with a romantic vision of nature (rural and anti-industrial) impregnated with nostalgia towards a virgin world (the idealized paradise). Likewise, in the 20th century Aldo Leopold became one of the first to articulate an ecological ethic arguing that every ethic that has been developed is based on the premise that the individual is a member of a community of interdependent parts. His ecological ethic or "Land ethic" widens these boundaries to include as part of the community elements such as soil, water, plants and animals, or collectively the Earth as a whole. Therefore, this "Land ethic" changes the role of *Homo sapiens* as a conqueror of the terrestrial community to a mere member and citizen of it implying respect for the other members as well as for the entire

© KONINKLIJKE BRILL NV, LEIDEN, 2022 | DOI:10.1163/9789004501270_013

community. However, as Leopold mentions the loss of paradise is related to the loss of an emotional-spiritual and material connection with the landscape or, in other words, it could also be related to the loss of affective ties with places.

To explain this link between landscape and affective ties and its relevance for this article, it is necessary to recall humanist geographer Yi-Fu Tuan and his concept of "topophilia." According to Tuan "topophilia" can be defined "broadly to include all of the human being's affective ties with the material environment" (*Topophilia* 93). Moreover, this can generate attachment to a place which helps nurture bonds between nature and its human community. In addition, Tuan described the importance of place for identity (Tuan *Space*, 4). Thus, in order for individuals to care about their environment there needs to be an attachment or bond to it. Along those lines, anthropologist Setha Low defines "place attachment" as "the symbolic relationship formed by people giving culturally shared emotional/affective meanings to a particular space or piece of land that provides the basis for the individual's and group's understanding of and relation to the environment" (Low and Altman 2). Therefore, a landscape's identity is linked to its inhabitants and the perception and relationship they have with it.

Drawing on these ideas, this article aims to address how the arts can contribute to making or re/creating a more sustainable society. In order to do so, the work of Lucía Loren and her experience with environmental art will be analysed. In particular, this chapter will address the following objectives. First, it will describe the relationship between humans and nature today, and how the arts can help educate our society in such times of crisis. Following this, it will focus on the important role of local communities and how change is initiated by the community's needs. To contextualize some of these topics, Lucía and I had a transatlantic conversation between Spain and the United States. We both come from different backgrounds: I am an academic interested in community building, social change and audience reception who has worked in the US and is now back in Spain while Lucía brings her insights as an artist and as an activist in Spain. Yet both of us are passionate about the role of the arts in environmental education and Lucía's art is a case in point. Thus, this article will analyse Lucía's work, her process, her interaction with the space and the community, and the short and long-term impact of her interventions.

2 Relationship between Humans and Nature

The General Conference of the United Nations Educational, Scientific and Cultural Organization meeting in Paris in 1972 established that the "deterioration or disappearance of any item of the cultural or natural heritage constitutes a harmful impoverishment of the heritage of all nations of the world" (par. 2). Moreover, it considered that, due to the magnitude and gravity of the new threats, "it is incumbent on the international community as a whole to participate in the protection of the cultural and natural heritage of outstanding universal value, by the granting of collective assistance which, although not taking the place of action by the State concerned, will serve as an efficient complement thereto" (UNESCO par. 2–3). In short, humankind, at all levels, should be working to find a solution and promote change.

While numbers show a fairly high percentage of concern among citizens in terms of environmental issues, it is still not enough. According to the Pew Research Centre survey of 2017 about 74% of US adults believed that the "country should do whatever it takes to protect the environment," compared to 23% that said the country has "gone too far in its efforts" to protect it (Anderson 2). Similar numbers have been released by the CIS (Centro de Investigaciones Sociológicas / Sociological Research Centre) in Spain, which showed that 76% of the citizens declared to be interested in ecology and the environment, compared to 24% that openly denied any interest (CIS 3–5). While these numbers show that environmental regulations seem to be more accepted today than they were before (albeit, especially in the US, it seems a question of political parties rather than true interest), it is also true that it has not "hit home" and therefore people do not care enough. In other words, although there has been considerable consensus regarding the environmental crisis, most people choose to look the other way. As critic José Parreño states, "such a high percentage of concern should generate a similar percentage of initiatives and commitments, but it is not the case. In conclusion, knowing the problem and saying you are worried about it does not mean working to remedy it" (111; my translation).[1] On that same note, Lucía in our conversation mentioned: "we live in a completely blind society which ignores the dramatic and irreversible consequences of what we are generating." As a result, one could question if the environmental crisis we experience today is really a crisis of consciousness. While talking to Lucía, she felt very strongly about the relevance of understanding the importance of an ecosocial vision in which we are all part of nature. This echoes UNESCO's insistence on considering the community as a whole to protect the environment.

As research—such as the one published by Nigel Nicholson (1998) and others—has confirmed, we are naturally wired to identify danger and to fix

---

1 All quotes from Spanish sources as well as the conversation with Lucía Loren are Sanchez-Vizcaíno's translation. This chapter is the result of numerous emails and conversations in Spanish between both authors, as well as background research by Sánchez-Vizcaíno.

problems, but what happens when society reads so many messages that people start wondering if they can really make a difference or not? The avalanche of messages and numbers result in psychic numbing (Slovic and Slovic 15). In addition, what happens when people ignore the threats and continue living without considering the consequences? According to Simon Thurley and his "Heritage Cycle," every process of finding and incorporating culture (and therefore in this case, nature) into our lives begins with understanding culture itself. Only when we understand it, we value it and by doing so, we learn to care for it, which will eventually help us enjoy it. The more enjoyment we get, the more we want to understand and care about it (Thurley 49). Hence, education and raising awareness seem the best way to generate understanding. However, when it comes to solving a crisis, we tend to believe in the sciences as the disciplines that will address the change. In our conversation, Lucía mentioned that she believed the only way of re-creating a more sustainable society would be from a third culture perspective in which humanities and sciences interacted to help us understand and assume the need of a radical shift in our way of life. In effect, all too often we have seen the ineffective use of numbers and statistics or as Scott Slovic and Paul Slovic call it "desensitizing effects of abstract quantitative discourse" (14). If we need people to understand, to connect, to value and care about nature we need "new modes of discourse—new ways of describing experience, new strategies for translating statistics into stories, new ways of articulating the meaning of both numbers and nerves. [....] Stories have the power to help us understand large, complex problems [...] that we cannot apprehend through quantitative information alone" (Slovic and Slovic 18).

3 Arts as an Educational Tool for Change

The arts might not be able to work alone but in Lucía's words they are capable of "communicating what is not visible to reason." She believes that the emotional impact is fundamental for deep understanding and to perceive the complex situation that our planet is suffering at this time. Maybe it is that emotional connection or maybe it is the power of storytelling that the arts have that presents to the audience the perspective of the Other (nature in this case). By perceiving the Other's need, the arts can instil the idea or feeling of empathy. According to several authors, the solution to the environmental crisis could be related to the rise of empathy from the community (Albelda and Sgaramella 14; Rifkin 605). I relate this to Lucía's concept of interdependence. According to Lucía, a key aspect to achieving a sustainable society is that everyone realizes the existing interdependence among living beings. We are all part of nature and part of a complex system that needs collaborative processes to create sustainable dynamics. If human beings do not seek a balance and understand that we live in a vulnerable system, they will not recognize that by exhausting the resources of this system we are exhausting our own existence. If this is not a crisis to solve and one that hits home, then what is?

The arts can help change a society or as bell hooks conceptualizes in her book *Art on My Mind: Visual Politics*, the arts have that transformative power that can resist and remove stereotypical ways of thinking. In this context understanding affect as an expansive term, alongside empathy, can be useful. For example, Melissa Gregg and Gregory Seigworth define affect as:

> (...) the name we give to those forces/visceral forces beneath, alongside, or generally other than conscious knowing, vital forces insisting beyond emotion—that can serve to drive us towards movement, toward thought and extension, that can likewise suspend us (as if in neutral) across a barely registering accretion of force-relations, or that can even leave us overwhelmed by the world's apparent intractability. (Loc 55)

Affect is defined here as emotional (automatic) responses that occur in relation to an object, person, idea, or activity beyond the social constructed responses. James Thompson's insight that "[a]ffect might produce a 'shock of thought' but that thought cannot be simplistically detached from the affective realm" (130) allows us to expand the meaning of this term. Thus, it is not only that emotional reaction we are seeking but consciousness-raising. Consequently, the arts can stimulate their creators and viewers combining pleasure and education in an absorbing moment "a moment that it is also potentially a propellant to a passionate commitment to social change" (Thompson 135).

On that note and given the recognition of arts as agents of change, recently, *Americans for the Arts* with *Animating Democracy* developed the Aesthetics Perspectives as a model to assess the impact of the arts. In doing research for this chapter, I found this model to be effective to analyse Lucía's work and its possible impact. This framework works as a reference to help artists, managers and scholars enhance understanding, describe and evaluate artistic and cultural processes geared "to progressive and positive change including justice, civic engagement, and community development" (Borstel and Korza 4). This framework describes eleven attributes that can help us analyse and assess the quality of artistic works: 1) "commitment", 2) "communal meaning", 3) "disruption", 4) "cultural integrity", 5) "emotional experience", 6) "sensory experience", 7) "risk-taking", 8) "openness", 9) "resourcefulness", 10) "coherence" and

11) "stickiness"[2] (Borstel and Korza1). Thus, as Susannah Laramie mentions, attributes such as openness, stickiness, sensory experience and emotional experience "can be used to assess the quality of artistic engagement in communities from a participant perspective" (6). Some of these attributes will be used to respond to how the arts can contribute to making or re/creating a more sustainable society through Lucía's work.

For the arts, one of the most important attributes listed in the Aesthetics Perspectives' framework is the "communal meaning." According to this framework, this attribute describes how the creative work can facilitate collective meaning "that transcends individual perspective and experience" (Borstel and Korza 17); in other words, highlighting how the work is relevant for the community. As mentioned before, the role of the community is fundamental to embrace change and to create a sustainable society. For instance, as Kristin Sedgman states in her article about audience reception studies, "while the first wave of audience studies asked what culture does to audiences, and the following iterations asked what audiences do with culture, audience research has more recently arrived at a third position, by asking how culture matters to audiences" (311). This brings us back to Thurley and his "Heritage Cycle" of what matters and that what we care about is what we value and enjoy. Many art projects in environmental issues have a basis in community. After all, and as Downing Cless puts forth, the "Greek root of 'eco'- is 'home.' Ecology (literally, "the study of home") promotes connectedness, rootedness, and holistic approaches to problems. Therefore, environmentalism is increasingly a grassroots movement: solving the earth's problems must start with community action […]" (80). In my conversation with Lucía, she linked the ideas of community to topophilia and to interdependence. According to the artist, sense of community is built on a strong identification of individuals and collectives with the physical space or territory. This identification promotes the emotional and affective bond with the inhabited place. Thus, art proposals can respond to the demands or conflicts of the place and its inhabitants since art favours the landscape interventions that derive from collective actions and interdependence with the rest of the biotic community.

4 Local Communities and Community Building

From a community building perspective, the arts can help empower community members and their abilities to work together toward resolving communal

2 "Stickiness" refers to the resonance, impact or value that a creative work achieves (Borstel and Korza 7).

issues. In his book *Bowling Alone*, Robert Putnam maintains that the core idea of social capital theory refers to the connections among individuals. Thus, as Lucía mentions, creative processes can facilitate this connection so that the link between collective stories, the custody of the territory and the need to solve problems becomes a significant environmental learning tool. This tool opens new communication channels that involve both the social and emotional levels, helping us understand the interconnection and interdependence of life and communities. As an example, Lucía believes that in rural areas there has always been a deeper sense of community, not only to survive but to achieve a "good life" and a sense of belonging and that is why human experiences should not be understood separately from the environment in which they occur.

The embeddedness of human experiences in landscape reaffirms community members as experts and not mere spectators (Sedgman 313), highlighting the importance of participatory models in the development of the arts. Paulo Freire asserts that by engaging the community in this creative dialogue we are encouraging them to develop a critical awareness of their own reality (conscientization) so they can act together in collaboration with the artist to transform that reality (action/reflection) (88). Marta López, Lucía Loren and Mario Vega reflect on this stating that:

> Community participation applied to the analysis of environmental problems and the search for solutions helps understand the complexity of these problems and to seek consensual solutions that take into account the needs of a broad spectrum of society, including the needs of nature itself. These solutions, therefore, achieve the dual objective of empowering the population and respecting life. (5)

Moreover, as José Albelda and Chiara Sgaramella discuss, this co-authorship from a collaborative and participative perspective is a unique way to enhance the process of empathy (17).

When I asked Lucía about this co-authorship she explained that when she works in the different communities there is a very close contact with the local population, which usually generates a shared learning: "I learn from them, from their cultural memory, from their way of working the land, and they learn to value their environment which is now 'dignified' as part of a visual narration that highlights many of the things they took for granted, especially in this industrialized and technological society that doesn't value traditions." Once more, this exemplifies the potential role of the arts as a tool for critical and creative reflection and the importance of engaging the community in the process. As Lucía comments "the change to a more sustainable society will only be

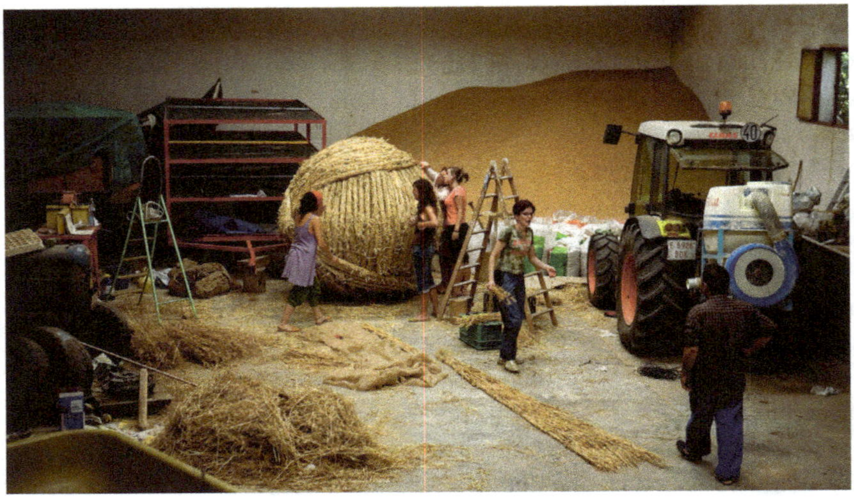

FIGURE 11.1 Al hilo del paisaje/ Landscape Thread. 2007 Paja y cuerda/Straw and rope. Arte en la Tierra/Art on Earth. Santa Lucía de Ocón, La Rioja.

possible from a collective effort." It is through these experiences how the community is empowered and undertakes collaboration and growth.

As an example of this empowerment Lucía shared with me one of her interventions in Santa Lucía de Ocón, "Al hilo del paisaje/Landscape Thread." This artistic intervention consisted in the creation of a large ball of thread (made of hay) whose "thread" traced a path to the cultivated field. The goal was to provoke a reflection on the relationship between cultivated space and "woven" space. The work was carried out with the participation of the community with whom she had spent a week weaving the ball with the same hay from the field where it would be later placed. Lucía explains that the sense of connection was very high and to illustrate this noted that on the last day some participants asked where in the field they should place "their ball." By calling it "their ball" they were expressing how they had made that work theirs and not hers exclusively.

5 Analysis of Lucía's Interventions/Interactions

The last example brings us to the third part of this chapter: the analysis of the quality of the creative process. This section focuses on Lucía's creative process and her background to look at ways in which her work illustrates a transition towards a more sustainable society. In order to do so, I asked Lucía several questions about her work and I analysed it using the Aesthetic Attributes mentioned before.

One of the first questions concerned her lifestyle and background and how that could be related to her work. Philosopher Fernando Arribas discusses the different approaches of Environmental Aesthetics and the difference between cognitivists and non-cognitivists (96). Of particular interest to the topic on hand was the differences between the scientific cognitivism (also known as the natural environmental model) from Allen Carlson and the integrated aesthetics of Emily Brady, and to what extent knowledge is necessary in order to value nature or if there are other alternative paths to understand, know and value it such as emotions and aesthetics. Thus, Lucía′s responses and descriptions of her lifestyle became relevant. When asked about her background, she explained where she lived, her concept of neo-rural lifestyle (those who studied in the city and decided to return to the countryside searching for life far away from industrialization), the conversations she holds with her neighbors or her passion to know and learn how life systems such as plants, animals, minerals, rivers or forests work. Lucía explained that knowledge was directly related with emotion and that emotion was the key element that makes the learning process meaningful: "this emotion arises from an inner need to link with what you are passionate about and amazed at." With this answer it seems as if cognitivist and non-cognitivists were united (through that third culture mentioned before) in search of that common goal: learn to value and to care for.

Another question was for Lucía to categorize her interactions. For example, Ernesto Domínguez Varela proposes a taxonomy to categorize art projects that are related to the environment. The four categories he introduces are: 1) Art with ecological themes; 2) Art with ecological architecture; 3) Art of ecological activism; 4) Art with ecological responsibility (57). Although Lucía claimed that her works fall into all these categories she said she identified more with the last one: "For me the priority is the responsibility that comes when choosing materials and ways to interact with the social environment and ecosystem." Lucía explained that her intention was to foster a reflection on cultural memory and the balance between human beings and landscapes. According to Lucía, artwork on display in a public space should not only be integrated into the space for which it has been specifically conceived (site-specific), but should also be based conceptually on the reflection of the place it is occupying. Thus, the work itself should be committed to renew and/or restore the perception of that landscape. This recalls Albelda and Sgaramella's analysis of "Hablando con Herbatium" by Lorena Lozano, a project which "allowed participants to rediscover the link of belonging to the territory and to the collective memory related to the plant world due to its site specific and experiential character" (20). It is this participatory and collaborative reflection what helps participants enjoy not only the aesthetics of the artwork but to engage emotionally with their own community and landscape.

Moreover, if Lucía's interview were to be word codified, one of the words that she frequently repeats is respect. According to her, it is this respect that underlies the ethic of care. This ethic of care understands the world as a relationship network; as ecofeminist philosopher Karen Warren argues we are defined by those relationships: "Relationships of humans to the nonhuman environment are, in part, constitutive of what it is to be a human" (33). Moreover, as philosopher Val Plumwood points out, the relationship with the land, the understanding of place rooted in memory and experiences is an essential ingredient in both personal and community identity formation (233). Place sensitivity implies listening and understanding the "language of the land" (Plumwood 231), something which Lucía works hard at before beginning any project. Plumwood highlights that we must be attentive to the "communicativity and intentionality of the more-than-human other" (230) to understand and value the land, something central to Lucía's process. Lucía emphasizes that every cultural, mental and mythical space determines the action or intervention. That is why before starting any intervention Lucía starts with the research of the place, its cultural memory, talking to the people, "reading" the environment and taking walks to feel the place. She begins to shape her idea from the material, human, emotional and symbolic resources while trying to perceive the need for social interaction the specific site seems to demand. By attentive listening, trying to understand the people´s experiences and by connecting this intervention to their cultural memory, she establishes this personal connection with the community, which she believes, will have a positive impact.

Most of Lucía's works have been defined as interventions. However, when asked, Lucía explained that if she could choose, she would change this pre-established term for interactions. In her opinion this last concept implies the open dialogue process with the environment, the inhabitants and the biotic community, something that does not always occur in an intervention. "Al hilo del paisaje/Landscape Thread" was an example of this open dialogue with the inhabitants as well as an example of participatory practices. In the same line, and as an illustration of the dialogue with the biotic community we have "Madre sal/Mother Salt." In this interaction mammals, such as domesticated goats, sheep, cows and horses or wild animals as roe deer, actively collaborated in it. The artwork consisted of burying in a meadow salt rocks carved as "breasts," so these animals could lick the surface to obtain the minerals they needed[3]

3 Salt in mammals facilitates their growth, prevents dehydration, and promotes digestion and assimilation of food, improving the health status of the animals that consume it.

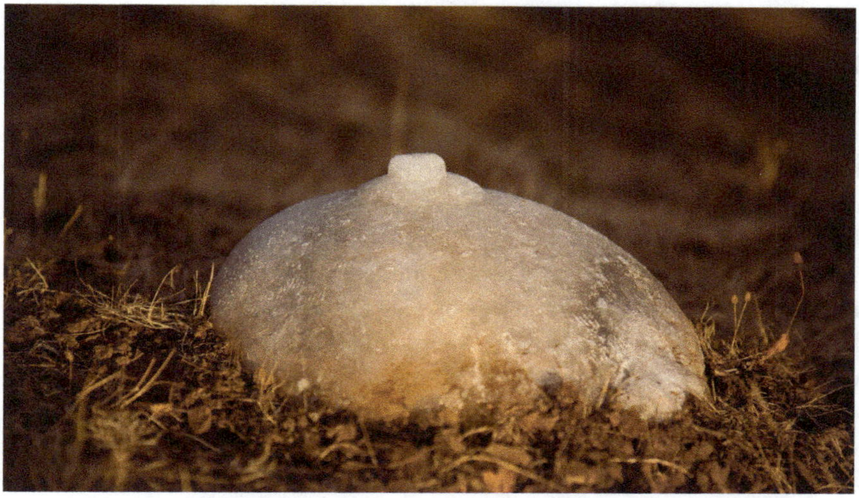

FIGURE 11.2 Madre sal/Mother Salt, 2008, Parque esculturas Lomos de Orio. Villoslada de Cameros, La Rioja.

until the sculptures disappeared from the landscape culminating the interaction. Taking into consideration the Aesthetic Attributes previously mentioned, one could argue that this project was at first characterized by "resourcefulness" since it used available resources and demonstrated a responsible social and environmental practice; in the end, there was no trace of Lucía's work. I would also highlight the component of "commitment" of the creative process to help the animals of the area. The creative work also assumed a risk ("risk-taking") not only due to the required active participation of the animals and the symbology behind the sculptures, but also the ephemerality of the work itself.

Another interaction that shows the open dialogue and participation of the community is the art creation "Bosque Hueco/Hollow Forest." In this interaction Lucía collected testimonies from the inhabitants. Their voices and stories created a "web of thoughts and feelings that described the importance of the environment." These stories highlighted how the trees form part of the community's identity and the need to care for them. In order to preserve them better, this intervention consisted in filling out and mending the hollows of four oaks and a centennial ash tree with branches collected from the environment. These plant structures protected the interior of the tree's "wound" creating a space of retention so that leaves and snow could not accelerate the process of putrefaction. The reflection of this work revolves around how changes in lifestyles and centralized environmental policies are leading to the disappearance of these landscapes. This work illustrates several of the Aesthetic Attributes. For example, the "cultural integrity" attribute as it demonstrates "the ethical use of material with specific cultural origins and context": in other words, how the artist respects the cultural memory of the environment, honouring the concerns of the inhabitants and working with them. In addition, the works responds to the attribute of "communal meaning" since it facilitates the relationship between stakeholders and explores the relevance of the landscape for this community by helping them to preserve it ("commitment"). Moreover, the work evokes the "emotional experience" of its participants by showing us a landscape that, like the trees, has been modelled and sculpted by the interaction of its inhabitants for hundreds of years. This emotional connection and the story collection part relates to the "openness" attribute, since the creative work remains subject to multiple influences, offering points of entry for people to connect with the work in different ways. Finally, it is important to highlight two more attributes. Once again, this work demonstrates Lucía's commitment to use materials that are available and that will reduce the environmental impact ("resourcefulness") and her intention and purpose of helping with her interactions ("coherence").

Lucía's work "Artesanía de un surco/Crafting a Furrow" also illustrates the importance of an open dialogue and the role of community members as experts. In this occasion, the shepherds themselves suggested that Lucía "sew" a furrow in order to avoid the erosion caused by the passage of cattle. Thus, she created a vegetal structure woven with wicker branches that covered the hollow of a gully and protected it from sun exposure and the impact of the rain to favour the growth of a vegetal cover inside: "a wicker oasis with green leaves woven in." This structure also facilitates the retention of earth and stones dragged by storms, filling the gap of the gully and helping to fix the structure of the soil. Consequently, this interaction serves as an example of the relevance of taking into consideration not only scientific knowledge but also the "traditional" knowledge that comes from experience and relationship with the environment. As Lucía states: "Reassessing and placing the spotlight on the importance of the shepherd's knowledge is fundamental, especially in order to reverse the idea or misconception of the rural as failure or poverty." In this way, and in terms of the Aesthetic Attributes, this project shows the relevance of the work for the community members and their collaboration ("communal meaning"); the "commitment" and "coherence" of the work by understanding a problem of the area and establishing a meaningful connection between the art piece and the land; "resourcefulness" by carrying out an intervention responsible with the environment, and last but not least, "cultural integrity" due to the use of materials and resources from that context.

Lucía is very conscious of the use of materials and the impact of her interactions: "In my projects I always try to minimize the ecological 'backpack' or

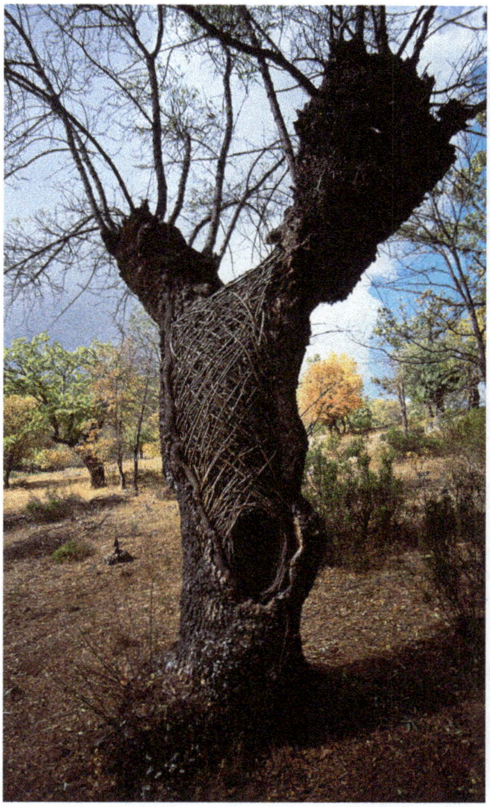

FIGURE 11.3
El bosque hueco/Hollow Forest, 2004.
Puebla de la Sierra, Madrid.

impact by using renewable and local resources, obtained in a respectful way, and avoiding the generation of waste that the environment cannot absorb itself." This is related to what Marco Ranieri mentions in his article when he states that "the use of an adequate poetics of materials in the process of artistic creation, allows us to situate ourselves in an attitude towards nature not only contemplative, but of total communion" (29). In other words, the use of materials from the land and their "adequate poetics" is fundamental to "generate the sensory stimuli necessary to arouse the feeling of empathy and can be favourable for the purposes of art in the genesis of an aesthetic of ecodevelopment, which contributes to the urgent change of our worldview" (Ranieri 36).

On that same note, I would like to comment on two more of Lucía's interactions, particularly due to the specific materials she has used. The first one is "?", part of the Carrícola Biodiversity Art Event in Valencia, an intervention in an abandoned agricultural plot, where marigolds were planted tracing a question mark in the field of cultivation. This plantation was used by artisans of

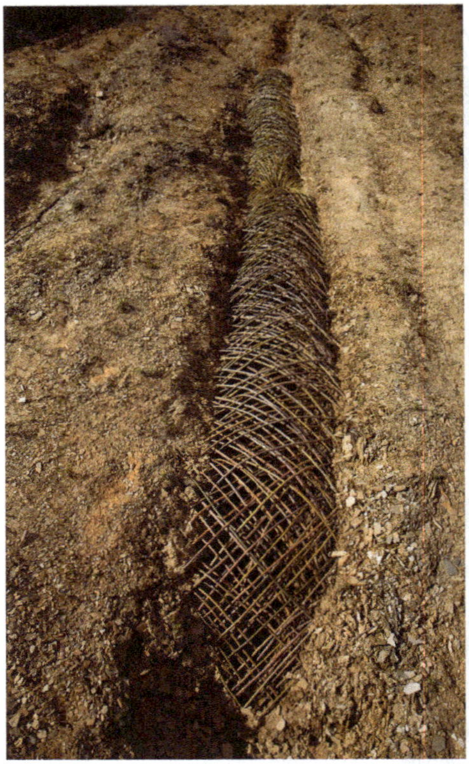

FIGURE 11.4
Artesanía de un surco/Crafting a Furrow. Mimbre y cárcava/Wicker and gully. 2009 Puebla de la Sierra, Madrid.

Carrícola and pollinating insects attracted by the flowers. The intervention opened with a performative action with fire around the interrogation, celebrating the spring equinox and the beginning of the agrarian cycle. Later calendula seeds and lavender plants were sown. This interaction shows the use of living plants in the creation processes that, as Ranieri assures, helps renew the bond of empathy between spectators and nature (29).

The second one is "Coser la cima/Mending the Top" where Lucía used wool to sew together the cracks of the parched soil on the top of a hill. The goal of this interaction was to show us the links between the animal and vegetal worlds as well as the process of soil regeneration with the arrival of rain and animal excrements, which would bring fertility to the soil. This interaction juxtaposes concepts such as erosion/fertility or deterioration/mending, creating strong emotional links.

Besides the use of sheep wool for the art piece, which brings us back to the idea of using sustainable materials, Juanma Valentín's audiovisual piece that combines this intervention with "Artesanía de un surco/Crafting a Furrow" deserves mentioning. While both interventions reach out to the local

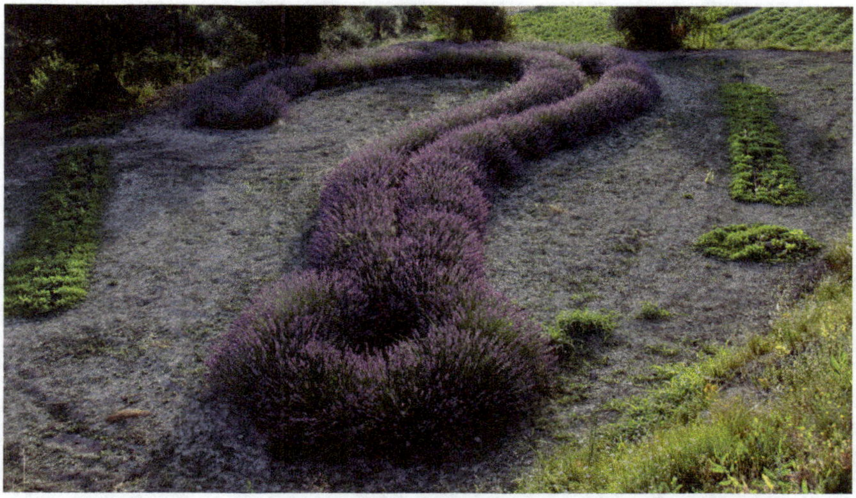

FIGURE 11.5 ? Caléndulas y lavanda/Calendulas and lavander. 2016 Biodivers, Carrícola, Valencia.

community, through these visual documentaries Lucía has found a way of disseminating her work to a broader community. Thus, the audiovisual documentary "En la piel/In the skin" reflects the "last skin" of the landscape, the deterioration of its vegetation cover and the erosive processes that are leading to a progressive desertification of soils. This audiovisual piece serves to illustrate one of the most important ecological problems currently in the Iberian Peninsula. In terms of the Aesthetic Attributes, this work is characterized by "openness." Thanks to the combination of the art piece and its visual documentary, access to the creative work is wider and far more people can interact with it. Hence, Lucía's work, with the collaboration of Juanma Valentín, has found a new way of spreading her message and raising awareness of the environmental issues that currently affect our country.

Last but not least, Lucía was asked about the impact of her work. She began highlighting once again the importance of reducing the natural impact of her work by using local resources and materials that would not generate waste and could be absorbed by the environment. She then talked about the "results" of some of her projects such as "Artesanía de un surco/Crafts of a Furrow" where, four years later, the wicker weaving had helped the accumulation of earth inside creating soil and slowing down the advance of the gully. Finally, she spoke of the importance of raising environmental awareness and consciousness through education and communication. One of her biggest concerns is how to reproduce and communicate some of her interactions and how to reach out to more people. As she mentions, environmental sustainability does

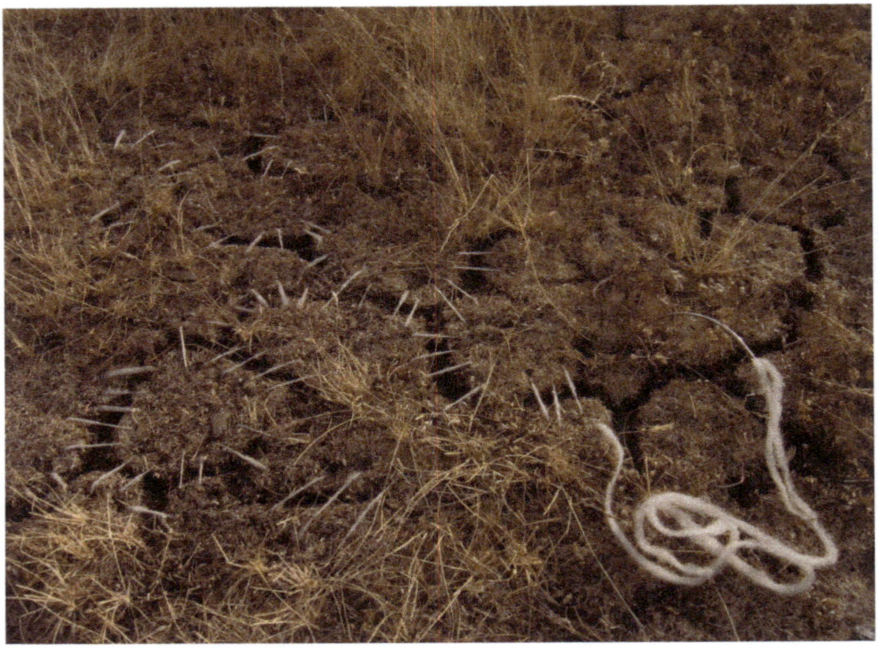

FIGURE 11.6 Coser la Cima/Mending the Top, 2007. Lana y grietas/Wool and cracks. Puebla de la Sierra, Madrid.

not depend on the individual and isolated actions; rather it needs social cohesion and collaboration to generate a significant change.

According to Lucía, this collaboration is a powerful resource to reactivate our social creativity with the effect of re-discovering a connection with the natural world. Thus, an education not only of scientific knowledge but of social and spiritual values is essential. For example, Maria Novo explains that in order to educate environmentally we need to develop a complex and critical way of thinking that does not oversimplify what was born "interwoven: life, nature, and our societies" (qtd García par. 13). Education, in the broad sense of the word, can open doors to both cognitive and emotional responses. Therefore, the spectator does not need to have any scientific knowledge or view art in a specific way. An artwork can excite and communicate in diverse ways and each viewer will interact with it in a completely different way. That is, in Lucía´s words, "the true value of an art piece: the layers that allow us to connect emotionally, in a sensorial way (including our body), rationally and of course spiritually."

Returning to the initial question of how the arts can contribute to making or re-creating a more sustainable society, philosopher Carmen Velayos, in an article on transitions and Deep Ecology, ends with some inspirational ideas among which are: cherishing life including the inert beings which are part of life and on which we also depend, re-evaluating our relationship with nature and rejecting human dominion over it and cultivating feelings and emotional access to nature (86–87). As I have shown, Lucía's work, as that of other artists, can help with this transition through her interactions.

To conclude, when asked whether it was worth carrying out these interventions, Lucía responded:

> I am not sure there is a clear answer. For me, it is a fundamental way to express my immense respect for "life" and the great frustration I have with the abusive use of beings, environments and materials. Therefore, yes, it is worth it. This type of work is worth doing because I do not know how I could live without getting involved in this problem and without communicating and raising awareness about it in some way.

Works Cited

Albelda, José, and Chiara Sgaramella. "Arte, empatía y sostenibilidad. Capacidad empática y conciencia ambiental en las prácticas contemporáneas de arte ecológico." *Ecozon@*, vol. 6, no. 2, 2015, pp. 10–25, ecozona.eu/article/view/662. Accessed 27 July 2018.

Anderson, Monica. "For Earth Day, Here's How Americans View Environmental Issues." *Pew Research Center*, www.pewresearch.org/fact-tank/2017/04/20/for-earth-day-heres-how-americans-view-environmental-issues/. Accessed 27 July 2018.

Arribas Herguedas, Fernando. "Belleza sin autoría: una visión de la apreciación estética de la naturaleza." *Ecozon@*, vol. 6, no. 2, 2015, pp. 94–109, ecozona.eu/article/view/668. Accessed 27 July 2018.

CIS (Centro de Investigaciones Sociológicas). "Barómetro de Mayo 2018. Avances de Resultados. Mayo 2018." datos.cis.es/pdf/Es3213mar_A.pdf. Accessed 27 July 2018.

Borstel, John, and Pam Korza. "Aesthetic Perspectives. American for the Arts: 2017." issuu.com/americans4arts/docs/aesthetic_perspectives_full_framewo. Accessed 27 July 2018.

Cless, Downing. "The Grassroots Is Greener." *The Drama Review* vol. 40 no. 2, 1996, pp. 79–102.

Domínguez Varela, Néstor. "Praxis y motivo: cómo y por qué hacer arte ecológico." *Ecozon@*, vol.6, no.2, 2015, pp. 54–66, ecozona.eu/article/view/665. Accessed 27 July 2018

Freire, Paulo. *Pedagogy of the Oppressed.* Bloomsbury Academic, 1993.

García, Leonor. "Entrevista a María Novo, Titular de la Cátedra UNESCO de Educación Ambiental y Desarrollo Sostenible de nuestra universidad." goo.gl/qi8bky. Accessed 27 July 2018.

Gregg, Melissa, and Gregory J. Seigworth. *The Affect Theory Reader*. Duke UP, 2010.

hooks, bell. *Art on My Mind: Visual Politics*. New Press, 1995.

Laramie, Susannah. "Companion Guide for Evaluators/Researchers." *American for the Arts*, edited by Pam Korza, 2017, issuu.com/americans4arts/docs/aesthetics_companion_evaluators. Accessed 27 July 2018.

Leopold, Aldo. *Una ética de la tierra*. Los libros de la Catarata, 2000.

Loren, Lucía. *Al hilo del paisaje* (Landscape Thread), Logroño, 2007. lucialoren.com.

Loren, Lucía. *Artesanía de un surco* (Crafting a Furrow), Puebla de la Sierra, 2009. lucialoren.com.

Loren, Lucía. *Bosque Hueco* (Hollow Forest), Puebla de la Sierra, 2005. lucialoren.com.

Loren, Lucía. ?, Biodivers Carricola, Valencia, 2016. lucialoren.com.

Loren, Lucía. *Carrícola Biodiversity*. Valencia, 2016. lucialoren.com.

Loren, Lucía. *Coser la cima* (Mending the Top), Puebla de la Sierra, 2009. lucialoren.com.

Loren, Lucía. *En la piel* (In the skin). Jaen, 2009. lucialoren.com.

Loren, Lucía. *Madre sal* (Mother Salt), 2009. lucialoren.com.

López, Marta, Mario Vega, and Lucía Loren. "El arte como herramienta para la educación ambiental." *Centro Nacional de Educación Ambiental*. 2017. www.mapama.gob.es/es/ceneam/articulos-de-opinion/2017-07-08-abril-vega-loren_tcm30-419306.pdf. Accessed 27 July 2018.

Low, Setha, and Irwin Altman. "Place Attachment – A Conceptual Enquiry." *Place Attachment*. Eds Irwin Altman and Setha Low. Plenum 1992. 1–12.

Nicholson, Nigel. "How Hardwired is Human Behavior?" *Harvard Business Review*, July-August, 1998, hbr.org/1998/07/how-hardwired-is-human-behavior. Accessed 31 Oct. 2018.

Parreño, José María. "Las artes del cambio." *Humanidades Ambientales: Pensamiento, arte y relatos para el siglo de la gran prueba*, edited by José Albelda, José M. Parreño, and José M. Marrero Henríquez. Catarata, 2018, pp. 109–24.

Plumwood, Val. *Environmental Culture: The Ecological Crisis of Reason*. Routledge, 2002.

Putnam, Robert. *Bowling Alone: The Collapse and Revival of American Community*. Simon and Schuster, 2000.

Ranieri, Marco. "Poética de los materiales en los procesos de vinculación empática. Significados y significantes de los materiales en la indagación artística sobre la relación Cultural/Naturaleza en el contexto de la crisis ecológica sistémica global." *Ecozon@*, vol.6, no.2, 2015, pp. 26–37, ecozona.eu/article/view/663/709. Accessed 27 July 2018.

Rifkin, Jeremy. *La civilización empática*. Paidós, 2010.

Sedgman, Kristin. "Audience Experience in an Anti-Expert Age: A Survey of Theatre Audience Research." *Theatre Research International*, vol. 42, no. 3, 2018, pp. 307–22.

Slovic, Scott, and Paul Slovic. "Numbers and Nerves: Toward an Effective Apprehension of Environmental Risk." *Whole Terrain. Reflective Environmental Practice.* Vol. 13: *Risk.* 2004–2005, pp. 14–18.

Thompson, James. *Performance Affects: Applied Theatre and the End of Effect.* Palgrave MacMillan, 2009.

Thurley, Simon. "Into the Future. Our Strategy for 2005–2010." *English Heritage: The First 21 Years. Conservation Bulletin,* Vol. 49. 2005, pp. 26–27.

Tuan, Yi Fu. *Topophilia: A Study of Environmental Perception, Attitudes, and Values.* Columbia UP, 1990.

Tuan, Yi Fu. *Space and Place: The Perspective of Experience.* U of Minnesota P, 1977.

UNESCO *Convention Concerning the Protection of the World Cultural and Natural Heritage.* whc.unesco.org/en/conventiontext/. Accessed 27 July 2018.

Velayos, Carmen. "Transición y ecología profunda." *Humanidades Ambientales: Pensamiento, arte y relatos para el siglo de la gran prueba,* edited by José Albelda, José M. Parreño, and José M. Marrero Henríquez. Catarata, 2018, pp. 71–89.

Warren, Karen. *Ecological Feminist Philosophies.* Indiana UP, 1996.

EPILOGUE

Chickens like Celebrities: A Short Story

José Manuel Marrero Henríquez

I had no doubts. I preferred chicken over pasta. I have nothing against pasta, in fact, I love noodles, but I preferred chicken, don't ask me why. It should have been a trivial decision, but it wasn't, and I swear that has made all the difference. Chicken, for God's sake.

It happened on my return flight from Chicago to Gran Canaria. I opened the plastic aluminum who-knows what box with anxiety. It's always the same. Nervous, desperate, I urgently unwrap the lunch box, even if I have already eaten a huge amount of the crappy food sold in airports: hamburgers, snacks, chocolate bars, candy. You're full but it feels like you're starving. It might be stress, or perhaps just the result of all that sugar and salt and preservatives that the body has been accumulating over hours of transit and connecting flights. Full but starving at the same time. Weird, isn't it?

It took me a while to reach the chicken hidden inside that goddamn box. I fought with the fold lid, burned my fingers, blew on them, tried again, burned myself several more times, and blew some more. I felt compelled to open it right away. Hungry or not, I didn't want to wait too long. Finally, I managed to take off the plastic aluminum who-knows what wrap and contemplated a few pieces of chicken drowning in a thick brown sea of sauce with mashed potatoes on the side.

Like Moses facing the Red Sea, I tried to divide that turbid liquid into two areas, chicken on the left and mashed potatoes on the right. But I was a lousy Moses and all I managed with that knife and fork was to leave the viscous fluid in the middle and put the chicken and potatoes on the uncontaminated shores. I quickly ate it all and left just a few tiny chunks floating in that disgusting fatty ocean. Then I had a small piece of brownie and drank some lemon soda. It was fast, faster than opening the box. I had lunch in a breath.

Sometimes you do things without thinking and afterwards you begin to think about what you've done. That's what happened to me on that flight from Chicago to Gran Canaria. Heavy digestion is not a good companion to thinking, but it keeps you awake and allows a couple of recurring ideas to go around and around. But it was more than ideas, it was the feeling of guilt, of having done something that should never have been done.

I felt like a bomb was lodged in my stomach and inside my brain. The chicken came to my mind and my neurons kept wondering. How long had it been dead before I ate it? Where was it born? What had its life been like? Why did I recall it as an it? Was it a "he" or a "she"? Perhaps none of the above. Let's say it was a "he," did he have a family? Wife, daughters, sons, parents, cousins?

He was the product of a massive artificial insemination program and conceived in a body that was already the product of various generations of artificial inseminations. A chicken family with no ties and no history. A non-family. Being male, his terrifying existential vacuum was not going to be as horrible as that of the hens. Thank God for industrial sexism, at least he wouldn't be raised in a box and forced to lay eggs all the time. He'd be raised in a concentration camp instead.

After hatching from his shell, he found himself surrounded by chicks running here and there and nowhere in the middle of an ocean of more chicks, jumping, shitting, sprinting, dashing from one side to another inside a huge space in an industrial farm. Uncomfortable, dizzy, anxious, scared of the noise, he was in a place that the crowd didn´t allow him to measure. He could feel the room, but he couldn´t see it. A tiny spot of yellow feathers in a storming sea of shouting companions.

People in white lab coats had checked his sex and put him together with a multitude of other male chicks. And there he was, in a smelly macho factory being fed to get fat for the rest of his life. Most of the time it was daylight, for there is no time to rest in the concentration camp for overweighting. Eating and drinking and growing, that´s all. Still, it was better than laying down to produce eggs until exhaustion and death.

He grew quickly and could always reach the food while many of his cellmates died from exhaustion. The elevation of the troughs increased automatically to follow a millimetric scientific pattern. He was lucky, good genes. Some could not deal with the permanent light and finally fainted to their death, some were not strong enough to stretch their necks and reach higher food. He did. He was strong, one more of that kind of strong chicken that ends up in the "high quality" menu of the lunch of a transatlantic flight.

Once big and fat he was sacrificed, though "sacrificed" is a little bit too much of a word for a lunch box product. No ceremony, no respect, he was just killed, industrially killed, and he didn´t even have time to feel that he was about to die. I'd say it was a less-than-heroic existence, even if there is no need for heroes in this dry epic tv show that life has become.

He had no opportunity of running on the fresh humid grass on rainy days, or strolling on the dry dusty ground on summer afternoons to clean his feathers. No sports, just getting fat to become a corpse to be eaten for the benefit and profit of the farmer. The chain of production and consumption that was his destiny from the very beginning, when he was just the potency of a spermatozoid in a test tube looking for an ovule. No New Testament. No Jesus. The meat industry was his Genesis and his Apocalypse all wrapped up together in a tidy plastic aluminum who-knows what box.

To pass away, as soon as possible, to be eaten and digested, to become fecal matter as soon as possible. No transcendent suffering, just torture that never stops, death and crime. Kill me. Do it quickly, please. No faith, no religion, no respect, no spiritual purpose, not even a brief memoir. No past, no present, and no future. The feeling of being feeling-less while having existed only to gain weight and be served on the menu of some transatlantic flight, just an alimentary bolus from the esophagus down to the gut with the help of a large amount of Coke or any other crappy soda, everything very domesticated and excessive, consumption, production, fleeing forward, the whole humanity fleeing forward in the chicken bolus.

No ecofriendly crap in the assembly line of the slaughterhouses of Chicago, clucking a wish out loud from a dark hole of nothingness at the end of the line, a voracious black hole that echoes the noise, the shouting, the countless fellow prisoners crying in the forgetfulness and void. It's the most beautiful thing you can get, a smelly black hole of perfect quiet unbeing, the end of misery.

There is not even a coffin of beans and carrots biologically cultivated, no healthy sauce where one's body can rest in peace; instead, just chemicals for a deathbed with flavorings of artificial flavor enhancers, palm oil, and every cheap shit that the market produces for the multitudes. That's no decent funeral. That's not polite. His dead body was mixed with chemicals and cooked and recooked until no substance of protein or vitamin was left, a chicken turned into a "filling thing" served with mashed potatoes made from potatoes previously dehydrated to be again hydrated, who knows why. Back and forth in a movement of suctioned water and repositioning that could never end well.

For God's sake, let's leave beings in peace, no fear of quietness, please, no fear of thinking, no hurry. Natural products should be slowly raised and cooked, their imperfections accepted, no need for that shining fruit in the supermarket that looks like a celebrity stretching her skin to maximum tautness, to the point of no return, eighty coming on sixty, but inside dust returning to dust. A simple chicken, please, raised in a country farm, not a Frankenstein-chicken recomposed, swimming in chemicals.

Chickens like celebrities,
fake like shining fruit.

His lying corpse upside down, his feathers removed, burned, and his body cleaned, the insides out and chopped off for further industrial processing, perhaps for burgers or fodder for the cows that later would become crazy, remember the mad cows? Vegetarians who ate the bones and remains of meat and entrails became crazy, and then some humans, after eating those crazy cows, also became crazy and that led to the killing of millions of cows, just to prevent further animal contamination.

The integrity of his body was not respected. He was chopped, diced and mixed with the pieces of other bodies. By chance some of his juiciest third-class muscles were selected for an airline menu. And there he was, or at least a part of him, two pieces that I ate with some brown sauce and mashed potatoes. I don't know what part of his body I chewed and I don't know if that especially hard piece was an extra dry portion of meat, his cartilage or another chicken's, or perhaps even a compact mixture of various specimens.

That box with chicken in sauce and mashed potatoes was a common grave, the microwaveable common grave that I ate for lunch while crossing the Atlantic Ocean. How many individuals were on my menu? One chicken, a part of one chicken? That might be too optimistic a thought. Let´s say I may have had between eight and ten bodies on my plate, then in my stomach, then in the fluids traversing my body, my blood, my saliva, in all my organs, in my brain, my lungs, even in my semen, just everywhere within me, and afterwards in my consciousness, just there bothering me for then and forever, even after that flight from Chicago to the Canary Islands.

I was no vegetarian, but I disliked the idea of having eaten several pieces of different chicken corpses who had lived miserable lives, all of them worse than the lives of the worst enslaved or kidnapped people. Yes, their suffering lives from their beginning to their end, all that was in my gut, revolving in a heavy digestion, mixed chicken cadavers going around and round together with the juices of my insides, candy bars, sodas, and other junk.

Back home, I could not get that chicken off my mind. He persisted in my stomach and in the long and heavy digestion that continued afterwards in my brain. It was not only a thought or a sentimental feeling, an abstraction, it was molecular, a matter that entered me and changed my life. That revolting mass of dead corpses in my stomach has made me more scrupulous. Now, I eat consciously. I must know what I eat before I eat it. I might be a better person now, but I have also become a pain in the ass, and my everyday life has turned into hell.

I had an easy life, well organized, no great responsibilities. I worked hard at times, but was never too engaged with anything. I had my Red Cross and International Amnesty memberships and a bad word for politicians from time to time, and that pretty much summed up my critical attitude with the world. But those dead chicken pieces on that flight from Chicago to the Canary Islands made everything different and changed my life. Fortunately, it has only affected just me because I don't have kids or a wife, and my parents have been dead for decades. I live by myself, for I have always been reluctant to any sentimental or filial attachment or commitment.

Before the flight from Chicago to the Canary Islands I took pleasure in the simple and trivial things of life. I could enjoy scratching my dandruff, searching for a snot deep inside my nose, capturing it, rolling it with my fingers and flicking it far away like a catapult, as far as possible. I loved defecating, especially when I was on the verge of wrecking my pants and arrived just in time to the bathroom, my insides exploding, at times in quite a nasty expansive way, leaving everything a mess. I could even enjoy its smell, as if I were a dog smelling the ass of another dog. Peeing was also a pleasure, smelling and sweating when I ran was also a pleasure, and drooling at night too.

Not anymore. At the beginning I was not aware of my new situation, I just felt uneasy in certain contexts. I felt goosebumps before going to the dentist, sometimes a week before the date, whenever I had to go for any kind of treatment, even a simple yearly tooth cleaning. I could not stand cutting my nails so I began to go to nail salons where women get their nails done with those ridiculous multicolor decorations. I went there just to have my nails cut, I didn´t need any of that hand beauty nonsense. I have ended up incorporating that habit in my monthly agenda. I feel stupid, but I still prefer having my nails done than cutting them myself. Except for my hair ---I am almost bald anyway, I refuse to deal with any excrescence of my body, nails, dandruff, mucus, excrements, scabs, even if they are mine and I live with them.

Those once-upon-a-time pleasures of everyday life have disappeared, and now I deal with them with extreme scrupulousness. I knew that something was about to change the day I went out with some friends for dinner. We went to a hamburger place, not one of those multinational chains, but a local place along the promenade of Las Canteras beach that offered good quality meat. I was distracted by the sunset and a beautiful girl playing beach-tennis on the shore when suddenly the dish caught my full attention. True, I had already had a couple of glasses of red wine from Lanzarote; but it was not the wine, it was the chicken from Chicago.

The idea of good quality meat revealed itself as an impossibility, an oxymoron, a misnomer, rubbish. I saw ground flesh and the corpses from where it had been extracted, for the menu clearly stated that the hamburger was made from a mix of cow and pork, of a not-clearly determined number of cows and of a not-clearly determined number of pigs, and that the piece of meat with cheese, onion and ketchup between two slices of bread was a crying animal holocaust.

If I ate that hamburger I would justify the millions of industrial killings taking place every day in the world, and that was something I did not want to justify. Each one of my hamburger bites would grind millions of bodies, cows, goats, horses, chickens, pigs, even dogs, cats, cockroaches and other animals missing in the Western habits of meat consumption. And with those animals I would be eating people, chewing those holocausts that fill human history everywhere in the world, in the Amazonia and in Germany, in the United States and in Bangladesh, in India and in the Belgian Congo.

It was not just because of my third glass of wine, it was the chicken from Chicago. I could not take my first bite of that burger. I handed it to my chubby friend Juan and ordered a plate of chickpeas. I had lived all my childhood in the country, on a banana plantation, surrounded by goats, cornfields, sugarcane, wheat, cows, hens, and that burger was also the foundation of my childhood, a childhood of rural feelings that I could not dare to chew for the sake of a frivolous dinner. I was not going to eat myself in an act of industrial suicide.

Needless to say, my life has become very difficult, miserable, to say the least. Here I am, on a tiny island, a little rock like a speck of dust in the Atlantic Ocean, back from Chicago, trying to live a simple life of natural respect. Everything seems to be against me. I am scared and lonely.

Names and Subjects Index